Measurement Uncertainty:
Methods and Applications
Fourth Edition

Measurement Uncertainty:
Methods and Applications
Fourth Edition

by
Ronald H. Dieck

Notice

The information presented in this publication is for the general education of the reader. Because neither the author nor the publisher have any control over the use of the information by the reader, both the author and the publisher disclaim any and all liability of any kind arising out of such use. The reader is expected to exercise sound professional judgment in using any of the information presented in a particular application.

Additionally, neither the author nor the publisher have investigated or considered the effect of any patents on the ability of the reader to use any of the information in a particular application. The reader is responsible for reviewing any possible patents that may affect any particular use of the information presented.

Any references to commercial products in the work are cited as examples only. Neither the author nor the publisher endorse any referenced commercial product. Any trademarks or tradenames referenced belong to the respective owner of the mark or name. Neither the author nor the publisher make any representation regarding the availability of any referenced commercial product at any time. The manufacturer's instructions on use of any commercial product must be followed at all times, even if in conflict with the information in this publication.

Copyright © 2007
ISA—The Instrumentation, Systems, and Automation Society
All rights reserved.

Printed in the United States of America.
10 9 8 7 6 5 4 3 2

ISBN-13: 978-1-55617- 915-0
ISBN-10: 1-55617-915-4

ISA
67 Alexander Drive, P.O. Box 12277
Research Triangle Park, NC 27709
www.isa.org

Library of Congress Cataloging-in-Publication Data in process.
Dieck, Ronald H.
   Measurement uncertainty : methods and applications / by Ronald H.
Dieck. -- 4th ed.
      p. cm.
   ISBN 1-55617-915-4
1. Mensuration. 2. Uncertainty. I. Title.
   T50.D53 2006
   530.8--dc22
                                        2006036563

TABLE OF CONTENTS

Dedication

The author dedicates this edition to his wife, Donna;
to his three children, Mark, Duane, and Heidi,
and to their growing families of grandchildren.

PREFACE

This fourth edition of *Measurement Uncertainty: Methods and Applications* is designed to provide an understanding of the importance of the role played by measurement uncertainty analysis in any test or experimental measurement process. No test data should be considered without knowing their uncertainty.

Examples and problems have been included to illustrate the principles and applications of measurement uncertainty analysis. Sections on the business impact of measurement uncertainty and the treatment of calibration uncertainty have been included. The very latest in measurement uncertainty technology is included in this revised edition. In addition, all the material herein is in full harmony with the ISO *Guide to the Expression of Uncertainty in* Measurement. Specifically, the terminology and symbols are in agreement with the US National Standard on Test Uncertainty and in harmony with the ISO Guide. The term "standard uncertainty" is now employed throughout to indicate that each uncertainty mentioned is one standard deviation of the average for a group of measurements. Thus, instead of reporting "random uncertainty" and "systematic uncertainty," this text now uses the terms "random standard uncertainty" and "systematic standard uncertainty."

This material will be useful to test engineers, process engineers, control engineers, researchers, plant supervisors, managers, executives, and all others who need a basic understanding of the assessment and impact of uncertainty in test and experimental measurements. In addition, technical school, college, and university students will find this course useful in gaining insight into the impact of errors in their measurements as well as estimating the effects of such errors with measurement uncertainty analysis.

Acknowledgments

Special appreciation is noted for the original and pioneering work of Dr. Robert Abernethy, whose early studies permitted the development of the universal uncertainty model fundamental to the methods of this text. My personal and most sincere gratitude remains that which goes to my Lord and Savior Jesus Christ, who saw fit to grant me the talents and abilities without which this writing would not have been possible at all. In Him there is no uncertainty, only the certainty of life everlasting.

Ron Dieck, 2006

Unit 1: Introduction and Overview

UNIT 1

Introduction and Overview

Welcome to *Measurement Uncertainty: Methods and Applications*. The first unit of this self-study program provides the information needed to proceed through the course.

Learning Objectives—When you have completed this unit you should:

    A.   Know the nature of material to be presented.

    B.   Understand the general organization of the course.

    C.   Know the course objectives.

1-1. Course Coverage

This course includes the basics of the measurement uncertainty model, the use of correlation, curve-fitting problems, probability plotting, combining results from different test methods, calibration errors, and uncertainty (error) propagation for both independent and dependent error sources. Extra attention is directed toward the problem of developing confidence in uncertainty analysis results and using measurement uncertainty to select instrumentation systems. Special emphasis on understanding is achieved through the working of numerous exercises. After completing this course, the student will be able to apply uncertainty analysis techniques to most experimental test problems in order to help achieve the test objectives more productively and at lower cost.

1-2. Purpose

The purpose of this book is to convey to the student a comprehensive knowledge of measurement uncertainty methods through documentation that is logically categorized and readily utilized in practical situations that confront an experimenter. Course structure is purposely divided into units that represent specific segments of measurement uncertainty methodology. Using this approach, the student will be able to proceed through the text, learning the methodology in an orderly fashion, and then return to specific topics as the need arises during experimental analysis.

The book contains numerous exercises intended to provide practical experience in working measurement uncertainty problems; several example calculations are presented to provide learning clarity for specific principles.

1-3.   Audience and Prerequisites

This course is intended for scientists, students, professors, and engineers who are interested in evaluating experimental measurement uncertainty.

The course is presented on the college level and presumes that the student has received two years' training in engineering or science, can handle rudimentary principles of calculus, and has a calculator or computer to work the examples and exercises.

The material should be useful to test engineers, senior technicians, first- and second-line supervisors, and engineers and scientists who are concerned with deriving meaningful conclusions from experimental data. It should also be useful to students and professors in technical schools, colleges, and universities who wish to gain some insight into the principles and practices of measurement uncertainty.

1-4.   Study Material

This textbook is the only study material required in this course and is designed as an independent, stand-alone textbook. It is uniquely and specifically structured for self-study. A list of suggested reading (Appendix A) provides additional references and study materials.

1-5.   Organization and Sequence

This book is divided into eight separate units. Unit 2 is designed to provide the student an introduction to the conceptual thinking and basic statistics required for measurement uncertainty analysis. Unit 3 provides the details of a measurement uncertainty statement, including the proper characterization of error sources and their combination into an uncertainty statement for a test result. Unit 4 outlines a step-by-step method for summarizing the effects of various error sources and their treatment both before a test and after a test is run. Unit 5 deals with several specific methods required for knowledgeable uncertainty analysis. Special attention is paid to uncertainty (error) propagation and determination of the effect of an error source on a test result. Unit 6 discusses the weighting of results by their uncertainty, which is a method for obtaining a test result more accurate than any of several independent measurements of the same result. Unit 7 outlines several practical application techniques for dealing with error data. These techniques are often needed during the course of computing measurement uncertainty. The book culminates with Unit 8, which offers a comprehensive treatment on the presentation of measurement uncertainty analysis results.

Each unit is designed in a consistent format with a set of specific learning objectives stated up front. Note these learning objectives carefully; the material that follows will teach to these objectives. The individual units often contain example problems to illustrate specific concepts. At the end of each unit, you will find student exercises to test your understanding of the material. The solutions to all exercises are given in Appendix I.

1-6.   Course Objectives/Learning Objectives

When you have completed this entire book, you should be able to:

A.   Describe the basic approach to evaluating measurement uncertainty.

B.   Converse comfortably using measurement uncertainty and fundamental statistical terminology.

C.   Complete a detailed measurement uncertainty analysis for any test or experiment, including defining the experimental setup, locating and estimating the magnitude of the random and systematic sources of error, propagating their uncertainties (errors) into the test result, calculating the test result measurement uncertainty, and presenting the uncertainty analysis in a compelling fashion.

D.   Explain the significance of a test's measurement uncertainty in order to better plan experiments and tests in a more cost-effective manner to yield test data more suitable to decision making.

1-7.   Course Length

The primary premise of this book is that students learn best if they proceed at their own pace. As a result, there will be significant variation in the amount of time taken by individual students to complete this book.

You are now ready to begin your in-depth study of measurement uncertainty. Please proceed to Unit 2.

Unit 2:
Fundamentals of Measurement Uncertainty Analysis

UNIT 2

Fundamentals of Measurement Uncertainty Analysis

This unit introduces measurement uncertainty analysis, including random error (or precision), standard deviation, and systematic error (or bias). Methods are given to estimate the magnitude of the effects of the random errors and systematic errors.

Learning Objectives—When you have completed this unit you should:

A. Understand the purpose of engineering, experimental, or test measurements.

B. Know the underlying principles of statistics as applied to measurement uncertainty analysis.

C. Be able to characterize errors and uncertainties into either random (precision) or systematic (bias).

D. Recognize the distinction between error and uncertainty.

2-1. The Purpose of Engineering, Experimental, or Test Measurements

The purpose of measurements is the same whether they be engineering, experimental, or test measurements: to numerically characterize the state or performance of a physical or chemical process. Properly understanding the data obtained from such measurements is crucial to applying the knowledge thereby gained. The pressure to use test data for decision making is often so great that there is a tendency to assume the data are correct, even to the point of almost never reporting an estimate of the measurement uncertainty with its test result.

It is important to note that every measurement ever made by every engineer or scientist has been in error and will be so in the future. There has never been and never will be a case when a person measured a variable and obtained the *true* value. Error is the difference between the measurement and the true value.

However, there are several circumstances under which a measurement value is considered to be true. The most common is that value which supports the preconceived result expected by the measurer. These kinds of data are frequently called "good data." Often the terms "right on," "nominal," "exact," "expected result," "astonishingly close" (my favorite), and so forth are applied to such test results. Conversely, "bad data" and "poor results" are terms often used to describe test results that do not support the action or conclusions wanted by the measurer. Each of those terms are subjective descriptions of the quality of test data or test results. They are ambiguous terms and as such should NEVER be used to describe the quality of a data set.

An objective, standardized method for describing and reporting the quality of test data or test results is available. This book presents such a method. Measurement uncertainty analysis is a numerical method for defining the potential error that exists in all data. The knowledge of the measurement uncertainty of a test result is as important as the result itself in characterizing the state or performance of a process. Test results should never be reported without also reporting their measurement uncertainty. No manager or process owner should take action based on test results with an undefined measurement uncertainty.

The purpose of engineering, experimental, or test measurements is to develop enough knowledge about a process so that informed decisions can be made. Since all measurements are in error, a method must be employed to define just how much in error the measurements might be, as that will certainly affect the decisions made. This book provides such a method—that is accepted throughout the world by organizations such as the American Society of Mechanical Engineers (ASME) (Ref. 1), ISA, the U.S. Air Force (Ref. 2), the Interagency Chemical Rocket Propulsion Group (ICRPG) (Ref. 3), the International Civil Aviation Organization (ICAO) (Ref. 4), the North Atlantic Treaty Organization (NATO) (Ref. 5), the International Standards Organization (ISO) (Ref. 6), and others.

2-2.   Measurement Error Definition

Early work by such visionaries as Eisenhart (Ref. 7), Klein and McClintock (Ref. 8), and others came to practical application and fruition with the work by Abernethy published in the U.S. Air Force handbook on measurement uncertainty (Ref. 2). That work was careful to divide error sources into two types: bias (or systematic) and precision (or random). The universally present third category, blunders (or mistakes), was assumed to be absent due to good engineering practice, and so will it be in this book. However, it should be recognized that a careful, comprehensive uncertainty analysis will frequently uncover blunders that need to be corrected before a viable measurement system can operate properly.

Measurement uncertainty analysis is a function of the measurement system. It is necessary to completely define the measurement system before proceeding with an uncertainty analysis. After that definition, error sources may be treated as either random (precision) or systematic (bias).

Random Error and Random Uncertainty

General

This section presents the basics of statistics as applied to measurement uncertainty. The first major error type considered is often called random error (and is sometimes called precision error).

Whenever a measurement is made, sources of random error will add a component to the result that is unknown but, with repeated measurements, changes in a random fashion. That is, the error component added to the second measurement is uncorrelated to that which has been added to the first measurement. So it is with each successive measurement. Each has a random error component added, but none of the components is correlated. The errors may be related in the sense that they come from the same distribution, but uncorrelated in the sense that errors in successive measurements cannot be forecast from a knowledge of errors in previous measurements.

Error and Uncertainty, Random and Systematic

In this text we will distinguish between the concepts of error and uncertainty as well as random and systematic. These concepts will be discussed several times to reinforce their meaning and application.

Error is defined as the true difference between the true value of the parameter being measured and the measurement obtained, sometimes called the measurand. Since we never know the true value, we never know the error. If we knew the error, we'd correct our data. That, in fact, is a major reason we calibrate. We trade the large unknown error we'd have without calibration for the expected small errors of the calibration process. Error, then, is unknown and unknowable.

A calibration factor can be expressed as +9.2°F, 1.4 psia, 25 lb/h, and the like after a calibration. These calibration factors are not the errors of the measurements. They are the corrections that are to be applied after the calibration. Do not confuse calibration corrections with the term "error." What is then needed is an estimate of the accuracy, or uncertainty, of the calibration factor quoted. The error in the calibration factor is unknown, but uncertainty estimates the limits of that error with some confidence.

Uncertainty then is an *estimate* of the limits to which we can expect an error to go, under a given set of conditions as part of the measurement process. Uncertainty is not error, only an estimate of its limits. It can be expressed as ±2.3°F, ±4.8 psia, ±76 lb/h, and the like. Note that each uncertainty is symmetrical in this case, although that is not a requirement, and that no value for correction is given. All that is known here, for example, is that the temperature is somewhere in the interval about the result, R, or $R \pm 2.3$°F. There is no correction for R, only an estimate of the limits about it in which the true value of R is expected to lie with some confidence.

There are random errors and random uncertainties. Random errors are values that affect test data in a random fashion from one reading to the next. Random error sources (random uncertainty sources) are those sources that cause scatter in the test results. This is always true.

Error sources (uncertainty sources) whose errors do not cause scatter in the test results are systematic sources and are estimated with systematic uncertainties. Systematic errors are constant for the duration of the test or experiment. Systematic uncertainty estimates the limits between which we'd expect the systematic errors to lie with some confidence.

Evaluating and combining uncertainties requires the basic statistics which follow.

Statistical Considerations

Random error components are drawn from a distribution of error that is a Gaussian distribution or normal distribution. That is, the error components originate from a distribution that is described by:

$$f(X) = \frac{1}{\sigma\sqrt{2\pi}} e^{-(X-\mu)^2/2\sigma^2} \tag{2-1}$$

where:

μ = the population average
σ = the population standard deviation
X = a population value
$f(X)$ = the frequency with which the value X occurs
e = the base of the natural logarithm

The value of σ is calculated for that distribution as follows:

$$\sigma = \lim_{N \to \infty} \left\{ [\textstyle\sum_i (X_i - \mu)^2 / N]^{1/2} \right\}$$

(2-2)

where:

$\sum$ = Throughout this text, the Greek capital sigma, $\sum$, will be used as the summation sign over the index shown. In Equation 2-2, the index is "i." In all other equations the summation index or indexes will be similarly obvious when they are not explicitly noted.

μ = the infinite population average

X_i = the ith data point extracted from the population

N = the number of data points used to calculate the standard deviation. Here, the number of data points in the population.

The term σ describes the scatter in the X values of the infinite population about its average, μ. It is also often called σ_X. The effect of μ, σ (or σ_X), and N can be seen in Figure 2-1.

Figure 2-1 is a smooth curve. Its height above the horizontal axis of X values represents the relative frequency that the infinite population has at that X value, or that X value's frequency of occurrence. Also illustrated is the area under the curve that is contained within the interval ($\mu \pm \sigma_X$). That area is ~68% of the total area under the curve. The interval ($\mu \pm 2\sigma_X$) contains ~95% of the area, and ($\mu \pm 3\sigma_X$) contains ~99.7% of the area. The area under the curve is equal to the fraction of the population with values between the limits that define the area.

However, an experimenter never has all the data in the infinite population, but rather only a sample of N data points with which to calculate the standard deviation. Nevertheless, for a data sample drawn from a Gaussian-normal distribution, the scatter in the data is characterized by the sample standard deviation, s_X:

$$s_X = \left[\frac{\sum_{i=1}^{N} (X_i - \overline{X})^2}{N-1} \right]^{\frac{1}{2}}$$

(2-3)

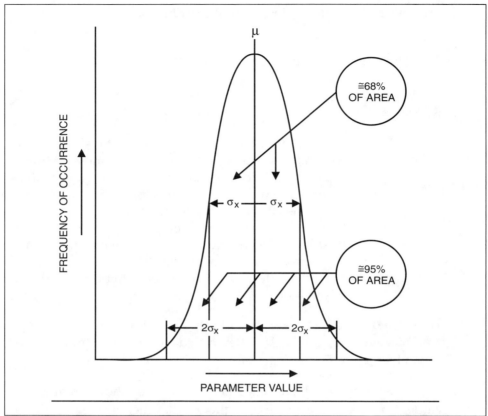

Figure 2-1. Gaussian-Normal Distribution

where:

$$X_i = \text{the value of the ith X in the sample}$$
$$\overline{X} = \text{the sample average}$$
$$(N-1) = \text{the degrees of freedom for the sample}$$

Degrees of freedom [sometimes shown as the Greek letter ν] will be discussed in detail later. Note here that uncertainty analysis uses the standard deviation and mean of a sample of data. The sample standard deviation, Equation 2-3, estimates the population standard deviation, Equation 2-2. In the same manner, the sample mean, $\overline{X}$, estimates the population mean, μ.

Equation 2-2 describes the scatter of an infinite set of data about the true population average, μ. Equation 2-3 describes the scatter of a data sample about its average, $\overline{X}$. Almost always, the uncertainty analyst will use Equation 2-3 because all the data in a population are seldom available.

It is worthwhile to note that computers usually have a code to calculate s_X, as do most scientific pocket calculators. A word of caution about the calculators: They often contain the capability to calculate both s_X and σ_X, which are sometimes noted as σ_{n-1} and σ_n, respectively. The experimenter always wants s_X, so watch out. To check on what your calculator does, use the two-point data set, 1 and 2. $\overline{X}$ should be 1.5 and s_X should be 0.707. If your calculator is calculating σ_X instead of s_X, you will get 0.5 instead of 0.707.

The effects of $\overline{X}$, s_X, and N are shown in Figure 2-2, which is analogous to Figure 2-1 but for a sample of data. Figure 2-2 is actually a histogram of a large data sample (i.e., more than 30 points). This plot is made so that the height of each bar represents the number of data points obtained in the interval shown by the width of the bar on the X-axis. Here the interval ($\overline{X} \pm s_X$) contains ~68% of the data sample, a situation similar to the infinite population but with $\overline{X}$ and s_X used rather than μ and σ_X. (This is also referred to as the 68% confidence interval.) Also, similarly, the interval ($\overline{X} \pm 2s_X$) contains ~95% of the data and ($\overline{X} \pm 3s_X$) contains ~99.7%. For uncertainty estimation, when there are fewer than 31 data points (i.e., less than 30 degrees of freedom), the factors of $1s_X$, $2s_X$, and $3s_X$ change because of the "t statistic," or Student's t. The student's t is a statistic that, along with the standard deviation (s_X), is used to determine confidence intervals (11).

For uncertainty estimation, when there are fewer than 30 degrees of freedom, one, two, and three s_X are not the correct multipliers to be used to obtain the aforementioned ~68%, ~95%, and ~99.7% confidence. What is simple but only approximately correct is to use ts_X, where t is Student's t. Appendix D contains the Student's t distribution for 95% confidence. Many texts (Ref. 9) on statistics contain Student's t for other confidences, but those tables are usually not needed for experimental uncertainty analysis. This text uses the notation "t" to mean "$t_{95,\nu}$" where the 95 is used to represent 95% confidence and where ν is the degrees of freedom.

It is necessary to use Student's t to compensate for the tendency to calculate sample standard deviations that are smaller than the population standard deviation when small samples of data are used. Here, small data samples mean fewer than 31 data points. Consider Figure 2-1; there is a great probability that a sample of only three or four data points will come from the center portion of the distribution, within ($\overline{X} \pm s_X$). This would result in the calculation of too small a standard deviation to properly represent the population as a whole. The same tendency is true of larger samples, although with decreasing likelihood of underestimating the

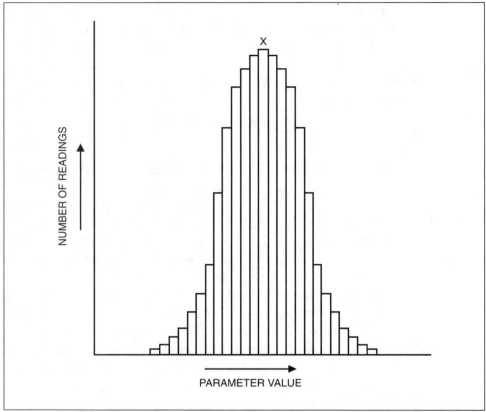

Figure 2-2. Gaussian-Normal Histogram

population standard deviation as the number of data points in the sample increases. Consulting the table in Appendix D, we note that once the degrees of freedom reach 30 (the number of data points reach 31), we can approximate Student's t with 2.0 with no significant loss in utility for any uncertainty statement. Student's t is also needed to properly predict the distribution of replicate averages, but that will be covered later.

One of the most elusive concepts in uncertainty analysis is a definition of "degrees of freedom" (d.f. or v). Texts in statistics are often either silent or presume too complex an understanding of statistics for the typical measurements specialist. Ref. 10 provides such a detailed discussion. This text defines "degrees of freedom" as the freedom left in a data set for error or variability. For each constant calculated from a data set, one degree of freedom is lost. (It is presumed that the data come from a normal random population.) Therefore, since the average, $\overline{X}$, is first calculated to obtain s_X, one constant has been calculated, $\overline{X}$, and there is a loss of one degree of freedom in the calculation of s_X. Sometimes $\overline{X}$ and s_X are called the first and second moments of the distribution of the data. For a detailed discussion of moments, consult Ref. 11.

This "loss of freedom for variability" is sometimes easier to visualize when one considers a straight-line fit to a data set, a subject to be discussed in some detail in Unit 7. For now, consider a two-point data set as shown in Figure 2-3. A line fit to two data points will of necessity go through both points exactly, since two points define a line. There is no freedom in the data set for any error in the line fit. The line hits both points exactly. Since a straight-line fit has two constants,

$$Y = aX + b \qquad\qquad (2\text{-}4)$$

two degrees of freedom are lost for each calculation of a straight line. Since there are only two data points, and since one degree of freedom is lost for each constant calculated, no degrees of freedom are left for error in the straight-line fit. The line goes exactly through the two data points.

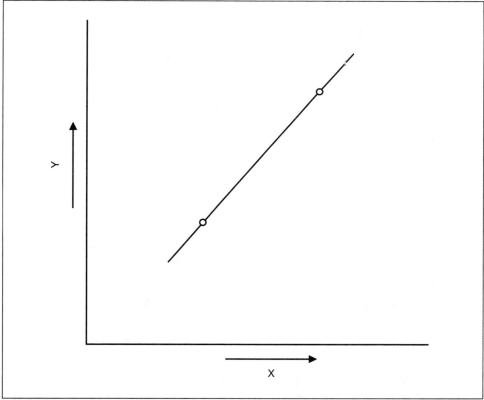

Figure 2-3. Two-Point Line Fit

Now consider Figure 2-4, where a straight line has been fit to three points. Note that not one of the data points is exactly on the line. Some error remains in describing the line. The degrees of freedom here is one (1): three data points minus two constants (or coefficients) calculated. The freedom left for error is shown in the scatter of the data about the line. Note that when there are degrees of freedom left, there is still room for random error to show up as departures from the calculated result—here, the straight line.

Remember, one degree of freedom is lost for each constant calculated from a data set.

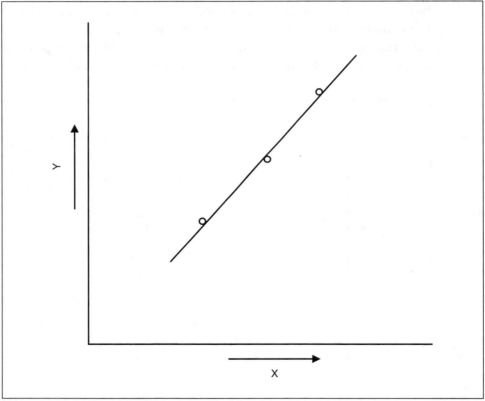

Figure 2-4 Three-Point Line Fit

Example 2-1:

To illustrate the computation and effect of $\overline{X}$, s_X, and ts_X, consider the following sample of readings taken from a 4 by 5 array of thermocouples in a duct of rectangular cross section. Note that here we assume that these data are normally distributed. If they were not, serious consequences would possibly result from applying standard statistical techniques of the type illustrated in this book. It is often advisable to check for normality with some standard technique, such as Bartlett's test.

For the data sample in Table 2-1:

$$\overline{X} = 100.16$$
$$s_X = 1.06$$
$$N = 20$$
$$\nu = 19$$

From Appendix D, we obtain $t = 2.093$.

Table 2-1. Thermocouple Data from Rectangular Duct

Reading Number	True Value, °F	Reading Value, °F	Error, °F
1	100.0	100.5	0.5
2	100.0	98.3	_1.7
3	100.0	99.0	_1.0
4	100.0	98.8	_1.2
5	100.0	102.4	2.4
6	100.0	101.3	1.3
7	100.0	100.6	0.6
8	100.0	99.6	_0.4
9	100.0	101.1	1.1
10	100.0	101.2	1.2
11	100.0	100.3	0.3
12	100.0	99.5	_0.5
13	100.0	100.3	0.3
14	100.0	99.9	_0.1
15	100.0	99.0	_1.0
16	100.0	99.7	_0.3
17	100.0	101.8	1.8
18	100.0	100.1	0.1
19	100.0	100.5	0.5
20	100.0	99.3	_0.7

The data in Table 2-1 have an average of 100.16°F and a standard deviation of 1.06°F. This means that, on average, approximately 95% of the data should be contained in the interval ($\overline{X} \pm ts_X$) = [100.16 ± (2.093 × 1.06)] = (100.16 ± 2.22), or from 102.38 to 97.94. (Remember that this is only an approximate interval. For a complete discussion of confidence intervals, tolerance intervals, and prediction intervals, see Ref. 12.) Reviewing the data we find that one data point, number 5, is outside those limits; 95% of the data is inside the interval described by ($\overline{X} \pm ts_X$). This is what we would expect. Amazing, it works!

A histogram of 5 segments can be constructed for the data in Table 2-1 as follows. (Note that here we are not trying to follow the standard methods for histogram construction, only illustrate a point.)

Choose the segments so that the low side of the bottom segment and the high side of the top segment include all the data for that segment. For 5 segments, divide the range of the data by 5 and round up to the nearest 0.1°F. The range of the data is the maximum minus the minimum, or $102.4 - 98.3 = 4.1$. Divide 4.1 by 5 to obtain 0.8^+ and round up to 0.9. Set up the segments, one centered on the average and two on each side of the average 100.16, as shown in Table 2-2, and note the number of data points in each segment, the frequency.

Table 2-2. Histogram Data

Segment Number	Segment Interval, °F	Frequency
1	97.91 – 98.81	2
2	98.81 – 99.71	6
3	99.71 – 100.61	7
4	100.61 – 101.51	3
5	101.51 – 102.41	2

Figure 2-5 is a histogram of the data in Table 2-1. Note that there are not many bars because we only had 20 data points. It still can be seen, however, that the shape approximates the normal curve in Figure 2-1. The 95% confidence interval approximation is noted as ($\overline{X} \pm ts_X$). The height of each bar is the frequency with which data in that interval occur. On a smooth curve for an infinite population as shown in Figure 2-1, the height of the curve is representative of the relative frequency with which that value shown on the X-axis occurs in the population.

Reviewing, we note that $\pm ts_X$ for a data set approximates the scatter of those data around their average, $\overline{X}$. s_X is the standard deviation of the data. $\pm ts_X$ is used with $\overline{X}$ to describe the interval, $\overline{X} \pm ts_X$, in which approximately 95% of the data will fall.

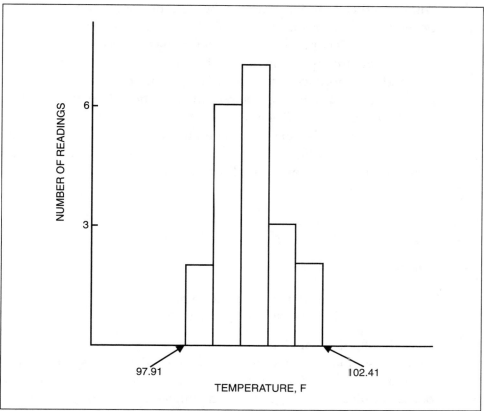

Figure 2-5. Histogram of Data in Table 2-1

Note that in Example 2-1 s_X has been used in a generic sense; no specific subscripts have been used to identify a particular error source. Also Table 2-1 has a column labeled "Error, °F." This is because we have stated the true value of 100°F. In practice, the true value is unknown as is the error.

The Standard Deviation of the Average, $s_{\overline{X}}$

Having now understood the calculation of the standard deviation and the utility of the value ts_X, it is important to note that usually a measurement is taken to understand the average of the set of data, not the spread or scatter in the data. What is usually needed or wanted is a statement about the error present in the average $\overline{X}$.

That desire may be expressed by asking, "How good is the average?" Since the term "good" is poorly defined conceptually and undefined numerically, it would be more proper to ask, "Were the experiment to be repeated numerous times, thus securing many averages, how much scatter would be expected in those averages?"

Although the second question is the right one, an investigator almost never has the opportunity to obtain the "many" averages. The next best question is, "Can the expected scatter in many averages be estimated from the data set already obtained with its one average?" The answer here is yes, and the approach is to utilize the central limit theorem. Ref. 13 provides an excellent discussion of the statistical derivation of the central limit theorem. It is sufficient for uncertainty analysis to note that the following is true: The expected standard deviation of many averages derived from an infinite population of a continuous, Gaussian random variable can be estimated as follows:

$$ s_{\overline{X}} = \frac{s_X}{\sqrt{N_{Ave}}} \tag{2-5} $$

where:

$s_{\overline{X}}$ = the standard deviation of the average, $\overline{X}$
N_{ave} = the number of X_i being averaged

Here the standard deviation of a set of averages is seen to be estimated from the standard deviation of the data, s_X. For the case in which the true population standard deviation is known, σ_X, the true standard deviation of the population average, $\sigma_{\overline{X}}$, is calculated as $\sigma_{\overline{X}} = \sigma/\sqrt{N}$. In the case of a data sample, however, s_X is how much scatter could be expected in a group of averages were they available from the population from which our one data set has been drawn.

Now the expression $ts_{\overline{X}}$ may be considered. $ts_{\overline{X}}$ can be thought of as "How good is the average obtained?" Or, 95% of the time, the population average will be contained within the interval ($\overline{X} \pm ts_{\overline{X}}$).

Note in Equation 2-5 that the number of data points in the reported average, N_{ave}, may be different from the number of data points used to calculate the standard deviation, s_X, N_{sd}. If s_X is from historical data with lots of data points (many degrees of freedom), N_{sd} will be very large (> 30). The number of data points in an average is N_{ave}, which is determined by the data sample averaged. (Sometimes, historical databases provide a better data set from which to calculate s_X than the sample averaged.)

s_X is often called the standard deviation of the average for a particular data set or sample. If there is only one error source, it is also the random uncertainty (formerly called precision error).

The impact of $ts_{\overline{X}}$ can be seen in Figure 2-6. While ~95% of the data will be contained in the interval $\overline{X} \pm 2s_X$ (for large sample sizes), it would also be expected that 95% of the time, the population average, μ, would be contained in the interval $\overline{X} \pm 2s_{\overline{X}}$. Note that, as shown in Figure 2-6, the expected scatter in X, s_X, is always larger than the expected scatter in $\overline{X}$, $s_{\overline{X}}$.

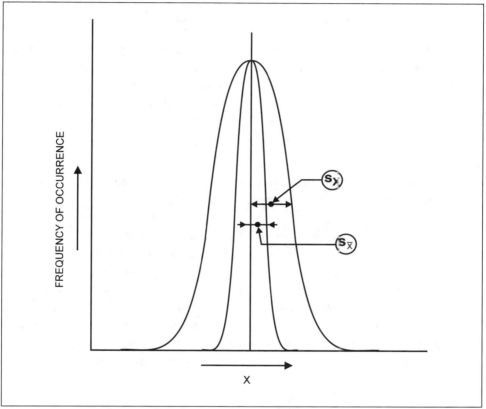

Figure 2-6. Normal Curve for Data, s_X; Normal Curve of Averages, $s_{\overline{X}}$

Recalling that the data in Table 2-1 yielded an $\overline{X} = 100.16$ and an $s_X = 1.06$ with 20 data points, the estimate of the standard deviation of the average, $s_{\overline{X}}$, is:

$$s_{\overline{X}} = s_X / \sqrt{N} = 1.06 / \sqrt{20} = 0.24 \qquad (2\text{-}6)$$

Here, the N is actually N_{sd}, which equals N_{ave}, which is 20. $s_{\overline{X}}$ is also called the random standard uncertainty. The term "standard" is used to indicate that this $s_{\overline{X}}$ represents an uncertainty source expressed as one standard deviation of the average. This important uncertainty concept will be dealt with in much more detail a little later.

From Equation 2-6 it is then noted that the interval $\overline{X} \pm ts_{\overline{X}}$ contains the true average for the population, μ, 95% of the time. That is, $[100.16 \pm (2.093 \times 0.24)]$ or 100.16 ± 0.50 should contain the true average of the population from which the data set was drawn. Since the data set in Table 2-1 came from an infinite population with $\mu = 100.$ and $\sigma_X = 1.0$, it is noted that the interval does indeed contain the true average for the population.

Since the interval contains the true average, it might also be stated that the true average temperature is in the interval quoted. In an actual experimental situation, it would not be known whether the true value was above or below the average obtained, only that it was contained in the stated interval of $\overline{X} \pm ts_{\overline{X}}$ with 95% confidence.

Pooling Standard Deviations

It should be noted that up to now only one estimate of the standard deviation has been discussed. There are times when several such estimates are available and one must consider which to use. Although it is obvious that the standard deviation with the most data is the most reliable, one can obtain a better estimate of the population standard deviation by combining several individual estimates of the same standard deviation through a process known as "pooling." Pooling several s_X values is done when each of them passes a homogeneity test, such as Bartlett's homogeneity test. Such a test ensures that each of the groups of data or samples comes from the same population. Assuming such a test has been made and passed, pooling may then be thought as averaging variances, $(s_X)^2$.

The equation used to pool s_X is:

$$s_{X,pooled} = \left[\frac{\sum\limits_{i=1}^{N} v_i \left(s_{X,i} \right)^2}{\sum\limits_{i=1}^{N} v_i} \right]^{\frac{1}{2}} \tag{2-7}$$

where $s_{X,\,pooled}$ = the pooled standard deviation and N is the number of standard deviations pooled.

The degrees of freedom associated with a pooled standard deviation is obtained as follows:

$$v_{pooled} = \{\sum(v_i)\} \tag{2-8}$$

where v_{pooled} = the degrees of freedom associated with the pooled standard deviation, $s_{X, pooled}$. It is just the sum of the degrees of freedom for all the individual estimates of s_X.

The several estimates of s_X may be from several days of data observed in a gas pipeline. Consider the three days' variability observed in the flow of natural gas in a pipeline as shown in Table 2-3. The best estimate of the standard deviation in the table, considering that no change in the standard deviation is expected between days, is that for day 2 because it has the most data. However, the pooled estimate is better than any individual day's estimate. It is calculated as follows:

$$s_{X,pooled} = \left[\frac{(5)(23)^2 + (11)(19)^2 + (8)(27)^2}{5+11+8} \right]^{\frac{1}{2}} = 22.8 \tag{2-9}$$

The best estimate of the population standard deviation is 22.8. The degrees of freedom associated with that standard deviation are:

$$v_{pooled} = 5 + 11 + 8 = 24 \tag{2-10}$$

Pooling standard deviations from several samples of data results in a better overall estimate of the population's standard deviation with more degrees of freedom than any one calculated standard deviation from one sample of data.

Table 2-3. Natural Gas Variability

Day	Standard Deviation, ft^3	v
1	23	5
2	19	11
3	27	8

s_X Using a Pair of Identical Instruments

Another method for estimating the standard deviation expected for an instrument or measurement system is to compare the readings taken at the same time on a process with identical instruments. Even if the process varies, as long as the instruments are observing the same process parameter (that is, the same level) at the same time, their difference can be used to infer the standard deviation of one instrument. s_X is computed as follows:

$$s_X = \left[\frac{\sum\limits_{i=1}^{N} (\Delta_i - \overline{\Delta})^2}{2(N-1)} \right]^{\frac{1}{2}} \tag{2-11}$$

where:

Δi = the ith difference between the two measurements

$\overline{\Delta}$ = the average difference between the two measurements

N = the number of i used to calculate s_X

When using Equation 2-11, only the differences between the two instruments are needed. The level does not enter the calculation.

Systematic Uncertainty

The Requirements for Accurate Measurements

When making a measurement, it is not sufficient to work toward small random errors, thus obtaining small standard deviations, for the data. An accurate result requires small random errors, but small random errors do not guarantee small measurement uncertainty, the basic requirement for accurate measurements. Figs. 2-7 to 2-10 illustrate that point.

Figure 2-7 shows a series of arrow strikes on a target where all the arrows have struck close together in the center of the target. There is low random error and low uncertainty. All the arrows are where they are supposed to be. The archer has been accurate. There is no systematic error of significance so the systematic uncertainty is also low. For a measurement system, this is the most desirable result and will be assumed to have occurred because of the low random error observed. We shall see later that this is a false sense of security.

Figure 2-8 illustrates the situation wherein the archer has scattered his arrow hits greatly but has still centered them on the target. There are significant random errors operating, but, on average, the accuracy is acceptable. There is still no significant systematic error. An experimenter will notice here the broad scatter, decide the measurement system is inadequate, and work to improve it. However, that decision will be made solely based on the scatter observed and not on the fact that unobservable systematic error also may exist.

Figure 2-7. Tight Arrows, Centered   Figure 2-8. Spread Arrows, Centered

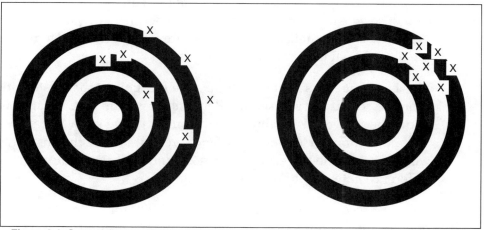

Figure 2-9. Spread Arrows, Off Center   Figure 2-10. Tight Arrows, Off Center

Figure 2-9 is the case in which the arrows are scattered far and wide and are well off the center of the target. Here the archer really needs help! The random errors are large and systematically to the right. There is some systematic error (or bias) that shows in the impact pattern of the arrows. In a measurement system, the experimenter will often recognize this situation as unacceptable, not because it is off-target but because there is so much scatter in the results. (Note that with modern data acquisition systems, the data may be rapidly averaged and only the average displayed. The experimenter will thus have no idea that there are large random errors operating.) Improvement in the experimental method is needed.

Figure 2-10 illustrates the arrows tightly grouped but off the center of the target. This is the most dangerous situation of all for an experimenter. The archer can see that the arrows are off-center; the experimenter cannot. The experimenter assumes the center of the target (true value for the measurement) is the center of the data (center of the arrows). There is low random error here, but this is not a sufficient condition to ensure low uncertainty. This archer (and, analogously, the experimenter) has significant systematic errors present and thus a significant systematic uncertainty should be estimated for this case. The archer can see it. The experimenter never can. The experimenter can see only the test data (the location of the arrows), not the true test result (the target). The experimenter will likely estimate the systematic uncertainty too low.

It is important now to note that an experimenter (or manager) would not be able in practice to tell the difference between the data set shown in Figure 2-7 and that shown in Figure 2-10. These two situations, which we see as different, would appear identical to them. Why? Because they cannot see the target when they run an experiment or test and secure data. We can see the target so we see that there is significant systematic error in the data shown in Figure 2-10.

This is a dangerous situation! It must be recognized that systematic error cannot be observed in test data. Remember, we run tests because we don't know the answer. (We cannot see the target.) If we knew the answer, we'd not spend money running tests. Since we don't know the answer, or since we cannot see the targets in Figures 2-7 and 2-10, we cannot tell the difference between those two data sets.

We must recognize that there is the possibility for systematic errors to operate and we must find a way to estimate their limits, the systematic uncertainty.

It is necessary, therefore, to include in uncertainty analysis the second major error type, systematic error.

Definition and Effect of Systematic Error or Bias

Systematic errors, formerly known as bias, are constant for the duration of the experiment. Systematic errors affect every measurement of a variable the same amount. It is not observable in the test data. The archer above could see there is systematic error in the aiming system because the target can be seen. However, remember that the experimenter does not see the target but assumes the center has been hit with the average of the measurements. Further, it is the insidious nature of systematic error that, when there is low random error, one assumes the measurement is accurate, with low uncertainty. This is not a sufficient condition for an accurate measurement. The random and systematic errors must both be low to have low uncertainty.

Figure 2-11 illustrates the effect of systematic and random errors. Since the true systematic error, β, is never known, its limits must be estimated with the systematic standard uncertainty, b. Note that in Figure 2-11 the scatter around the average displaced by systematic error is shown to be illustrated by σ_X. The difference between the displaced average and the true value is β. As b is an estimator of the limits of β, we note that β is contained in the interval $-b \leq \beta \leq +b$; so, the systematic standard uncertainty is expressed as $\pm b$. Note that the systematic uncertainty, b, may or may not be symmetric about the experimental average. In cases where it is not, it is expressed as $-b^-$ and $+b^+$. This will be discussed in more detail in Unit 3.

The systematic standard uncertainty, b, is assumed to have infinite degrees of freedom (in the absence of additional information). As such, it represents a 68% confidence interval, not 95% confidence. This is important. Once we get to estimating the systematic uncertainty, we note that it is usually the case that such estimates are approximately 95% confidence. This would mean that, with infinite degrees of freedom, Student's t would be 2 and we would divide our estimate of systematic uncertainty by 2.00 in order to obtain the systematic standard uncertainty, b. If the systematic uncertainty were obtained with degrees of freedom less than 30 (from calibration data for instance), we would divide that 95% confidence interval by the appropriate Student's t to obtain the systematic standard uncertainty, b. We will discuss this again later.

Five Types of Systematic Errors

Of the five types of systematic errors only two give rise to systematic uncertainties. These are summarized in Table 2-4.

Systematic errors of type 1 are large errors known to exist, and an experimenter calibrates them out. Examples include pressure transducer calibrations, thermocouple calibrations, voltmeter calibrations, and the like. In each case, it is known that data of unacceptably large error will be obtained from such an instrument unless it is calibrated. The calibration

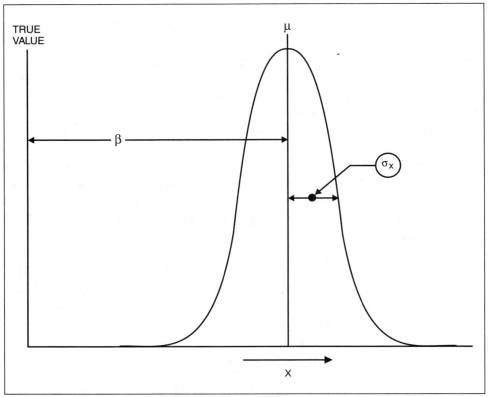

Figure 2-11. Normal Curve Displaced from the True Value

Table 2-4. Five Types of Systematic Errors

Size	Known Sign and Magnitude	Unknown Magnitude
Large	1. Calibrated out	3. Assumed to be eliminated
Small	2. Negligible contribution to systematic uncertainty	4. Unknown sign 5. Known sign

process trades the large error of not calibrating for the (assumed) small errors of the calibration process.

Systematic errors of type 2 are known to exist but are assessed as being small enough that they do not influence the outcome of the test measurement. One example might be the thermal conductivity of a thermocouple in a process stream. Such conductivity-caused systematic error may be deemed negligible for the test in question. If those errors are deemed negligible, their uncertainty estimates do not enter an uncertainty analysis. However, be careful here as errors assumed to be negligible are not always so. Prudent judgment is needed.

Systematic errors of type 3 are large errors that are assumed to be properly engineered out of the experiment or test. An example might be the proper use of flow straighteners in front of a turbine meter. Without those straighteners, swirl in the stream will cause the meters to read in error. The proper engineering practice of setting up the test, instrumentation, and data recording is deemed adequate to eliminate errors of this type. It should be noted that although these systematic errors do not have a part in uncertainty analysis, if they still exist in the measurement system, their presence often will be made known by the existence of data that are difficult to explain. These types of errors are often called "blunders."

Systematic errors of types 4 and 5 give rise to uncertainty estimates that are included in an uncertainty analysis. These systematic errors are of unknown magnitude but are not eliminated by engineering practice, not corrected out by calibration, and not negligible. If their signs are not known, they are expressed by the systematic standard uncertainty $\pm b$ in the uncertainty analysis. If the sign of the error has a known tendency, either positive or negative, it is necessary in uncertainty analysis to carry through two, nonsymetric, systematic standard uncertainties, $-b^-$ and $+b^+$. These represent the positive and negative estimates of the limits of a particular systematic error source. Nonsymmetrical systematic standard uncertainties will be detailed in Unit 3.

Obtaining Estimates of Systematic Standard Uncertainties

Since systematic errors are not known but only estimated, the term used to describe the limit likely for a systematic error is "systematic standard uncertainty." Coming up with an estimate of the magnitude for a systematic uncertainty source is often difficult. Ref. 2 provides five methods for obtaining systematic standard uncertainties.

Method 1: Tests with artifacts or samples run at several laboratories or facilities will do a fine job in estimating systematic error between facilities. The scatter in those data, usually expressed as $\pm ts_X$, can be used to express the systematic uncertainty for one facility or laboratory. This is because that one facility is assumed to be constantly different from the group whose spread is $\pm ts_X$, where X represents the several facilities or laboratories whose measurements are compared. The facility cannot correct its data to the average since, in a properly run intercomparison, facilities are not identified. If they are identified, corrections should still not be applied, as the spread is real evidence of systematic errors in the method and affects each facility.

Method 2: Where possible, an estimate of an instrument's systematic uncertainty can be obtained by transporting a calibration standard to the instrument in its operating environment. Numerous comparisons of this type will yield an estimate of the typical systematic uncertainty for that instrument in its application. These tests could also be used to calibrate the instrument and remove that systematic error source if so desired. It is important to remember, however, that continued in-place calibration is required to keep that error source eliminated.

Method 3: There are times when several independent methods are available to measure the same thing. For instance, jet engine airflow can be measured with the inlet bellmouth (nozzle), the compressor speed-flow map, the turbine flow parameter, and the exhaust nozzle discharge coefficient. Each of these may be configured to be independent measurements. The differences observed (in the average) are evidence of the systematic errors of the methods. The systematic uncertainty for any one method may be estimated by $\pm ts_X$, as in method 1 of Table 2-4, where X represents the several measurement methods.

Method 4: When it is known that there are specific causes for systematic error, special calibrations can be run to eliminate that error source. Failing that, repeated calibrations can estimate its magnitude for inclusion in the uncertainty analysis.

Method 5: When all else fails, one can use the judgment of several competent experimenters to estimate the systematic uncertainties. Beware of pride here. Many experimenters will start out by saying that their experiments have no systematic errors and, thus, their systematic uncertainty is, by definition, zero. This is never true and a reasonable, careful, thoughtful estimate can be had. In addition, one should use instrument manufacturers' literature with care as it represents the judgments of those producing the device for sale.

2-3.   Summary

Besides estimating the limits of random errors with random uncertainty, the limits of systematic errors must be estimated with systematic uncertainties for a complete uncertainty analysis. It is difficult but necessary. Their combination into uncertainty will be covered later.

References

1. ANSI/ASME PTC 19.1-2006, *Instruments and Apparatus, Part 1, Test Uncertainty*.

2. Abernethy, R.B., et al., 1973. *Handbook—Gas Turbine Measurement Uncertainty*, AEDCTR-73-5, Arnold AFB, TN: Arnold Engineering Development Center.

3. *ICRPG Handbook for Estimating the Uncertainty in Measurements Made with Liquid Propellant Rocket Engine Systems*, Chemical Propulsion Information Agency, No. 180, 30 April 1969.

4. ISO/DIS 7066-1.2, 1988-04-28, *Assessment of Uncertainty in the Calibration and Use of Flow Measurement Devices, Part 1, Linear Calibration Relationships*; ISO 7066-2, 1988-07-01, Part 2, Nonlinear Calibration Relationships.

5. *Recommended Practice for Measurement of Gas Path Pressures and Temperatures for Performance Assessment of Turbine Engines and Components*, NATO AGARD Advisory Report No. 245, June 1990, pp. 35–41.

6. *Guide to the Expression of Uncertainty in Measurement*, 1993. Geneva, Switzerland: International Standards Organization.

7. Eisenhart, C., 1963. "Realistic Evaluation of the Precision and Accuracy of Instrument Calibration Systems." *Journal of Research of the NBS 67C*, 2:xx–xx.

8. Klein, S. J., and McClintock, F. A., 1973, "Describing Uncertainty in Single-Sample Experiments." *Mechanical Engineering 75*, 1:3–8.

9. Bowker, A. H., and Lieberman, E. H., 1969. *Experimental Statistics*, p. 603. Englewood Cliffs, NJ: Prentice-Hall.

10. Ibid., pp. 114–116.

11. Hicks, Charles R., *Fundamental Concepts in Statistics*, 2nd Edition, Holt-Rinehart-Winston, 1964, p. 14.

12. Mendenhall, W., and Scheaffer, R. L., 1973. *Mathematical Statistics with Applications*, pp. 299–302. Doxbury Press.

13. Hahn, G., 1970. "Understanding Statistical Intervals." *Industrial Engineering* no. 12:xx–xx.

14. Mendenhall and Scheaffer, pp. 252–256.

Exercises:

2-1.   Student's t Problems

a. Write the expression for the standard deviation of the average as a function of the standard deviation of the data.

b. Write the expression for the random standard uncertainty as a function of the standard deviation of the data.

c. In your own words, describe what the equation of (b) above means with reference to replicate samples of $\overline{X}$.

d. Given $\overline{X} = 4.9$, $s_X = 1.7$, d.f. $= 11$

(1) Calculate the random component of the uncertainty (Student's t times $s_{\overline{X}}$).

(2) Write the interval that will contain μ, the true average, 95% of the time.

2-2.   Pooling Problems

Assume appropriate homogeneity tests are met.
Consider the following five estimates of the same σ:

s_X	325	297	301	280	291
υ	15	12	22	3	7

a. What is the best estimate of σ in the above?

b. Calculate the random error component of the uncertainty (t^*(random standard uncertainty) of the average or result) for each case.

c. A pooled s_X is better than any one s_X. Calculate $s_{X,\,pooled}$ and υ_{pooled}.

2-3.   Dependent Calibration Problems

A gasoline refinery decides to add a second volumetric flowmeter in series to each of the input fluid lines and output fluid lines in an effort to improve accuracy (reduce uncertainty).

a. Should the flowmeters be calibrated simultaneously or separately?

b. How would you decide whether an independent metrology laboratory should be contracted to calibrate one or both meters?

c. Only one determination of fluid density is available for each pair of meters. What effect does this have on the two measured weight flows in each line?

d. What is the best estimate of the flow through the refinery if we assume no losses?

Unit 3:
The Measurement
Uncertainty Model

UNIT 3

The Measurement Uncertainty Model

In this unit, procedures and models will be developed for providing decision makers with a clear, unambiguous statement of the accuracy or uncertainty of their data. No manager should ever be without a statement of measurement uncertainty attendant to each piece of data on which decisions are based. No experimenter should permit management to consider measurement data without also considering its measurement uncertainty. The manager has the responsibility for requiring measurement uncertainty statements. The experimenter has the responsibility for never reporting test results without also reporting their measurement uncertainty.

Learning Objectives—When you have completed this unit you should:

A. Understand the need for a single-valued uncertainty statement.

B. Be able to characterize error sources as systematic or random.

C. Be able to combine elemental systematic and elemental random uncertainties into the systematic uncertainty and random uncertainty for the measurement process.

D. Be able to combine the process systematic and random uncertainties into a single value of measurement uncertainty for the measurement process.

3-1. The Statement of Measurement Uncertainty

The purpose of measurements is to numerically characterize the state or performance of a physical process. Properly understanding the data from measurements requires a statement of uncertainty.

In the previous unit, the concepts of systematic error (and uncertainty) and random error (and uncertainty) were developed. It is not enough to be able to state the systematic and random uncertainties of a measurement. Management decision makers do *not* want to hear a flow level is 3000 scfm with ±30 scfm systematic uncertainty, ±14 scfm standard deviation of the average with 75 degrees of freedom. They want a single number from which to understand the limits of accuracy for the 3000 scfm measurement. That single number is the statement of *measurement uncertainty*. It is a single, unambiguous number. It is an objective estimate of the data quality.

3-2.   Grouping and Categorizing Error Sources

To conduct a proper uncertainty analysis, it is necessary to identify the error sources that affect an experimental result and to characterize them as systematic or random. ISO's Guide (Ref. 1) recommends grouping error sources and uncertainties according to the origin of their estimation. The Guide recommends using two types of uncertainties: Type A, for which there is data to calculate a standard deviation, and Type B, for which there is not.

This kind of Type A and Type B characterization is not required for uncertainty analysis. However, it is recommended that the uncertainty "type" as noted by ISO be used. The easiest way to indicate the origin of the uncertainty estimating data is to add a subscript to each elemental uncertainty estimate. In this way, the engineering grouping of systematic and random may be retained as well as the ISO grouping by information source. This compromise is recommended in the new, national standard on measurement uncertainty (Ref. 2).

In this book, where not otherwise noted, it will be assumed that all systematic uncertainties are Type B and all random ones are Type A.

In addition to the above, the bookkeeping process is eased when the error sources are grouped into categories (Ref. 3). It is instructive to group error sources into four categories: calibration, data acquisition, data reduction, and errors of method. However, this grouping of error sources (also uncertainty estimates) is not needed for an uncertainty analysis.

Grouping Error Sources

Calibration errors are those that result from the laboratory certification of an instrument response. Usually, this category includes errors that result from the use of instruments at the test site (sometimes called installation errors). It is very important to note that these are not the usual calibration errors that are calibrated out by the calibration process. For example, it may be known that a pressure transducer will have an uncertainty of ±2% if it is not calibrated and ±0.5% when it is. In this case, the large known systematic uncertainty (±2%) has been traded for the smaller uncertainties in the calibration process (±0.5%). Uncertainty sources totaling ±1.5% have been removed. What remains is the unknown error of the calibration process (which is estimated by an uncertainty). Only its limits are known:

±0.5%. It is this ±0.5% that will enter the uncertainty analysis (properly split between systematic and random uncertainty). These uncertainties are determined by a careful uncertainty analysis of the instrument calibration and use process.

Data acquisition errors are those that result from the use of data acquisition equipment, which may be computer data acquisition systems, a data logger, or just a person with pencil and paper reading a meter. Acquiring data to estimate the magnitude of these error sources is usually simple. All that is needed is to take multiple readings over the appropriate time period. This category of errors is also usually smaller in magnitude than calibration errors by a factor of three to ten. As with calibration errors, the exact magnitude of these errors is not known; only their limits may be estimated: systematic and random uncertainties.

Data reduction errors are those that most often result from the use of computers or pocket calculators. Truncation, roundoff, and approximate solutions are examples in this category. These error sources are usually smaller than either of the former two sources, but they can be significant— most commonly due to improper curve fitting. This will be discussed in Unit 7.

Errors of method have to do with the ability of the experimenter to adjust instruments on the test site. These errors are sometimes called personal errors. Sometimes this group also includes sampling error, that is, how well a measurement system is characterized considering its own variability or profiles. Sampling error is discussed in Unit 7.

Grouping errors into categories is not a necessary step to a correct uncertainty analysis. It is only a bookkeeping aid. (It is fully correct to conduct an uncertainty analysis with only one big group for all the error sources, as is done in this text.)

Note: Units 3 and 4 present the calculation of measurement uncertainty for a single measurement or parameter, such as temperature or pressure. When these measurements are then used to calculate a result, such as flow, the systematic and random uncertainties must be propagated into that result, using the methods of Unit 5.

Categorizing Errors and/or Uncertainties

Early in an uncertainty analysis it is necessary to decide whether an error source is systematic or random. The defined measurement system will provide the information necessary to make this choice. There are times when the same error source is a systematic error in one circumstance and a random one in another.

For example, in a multifacility test, the scatter calculated between facilities would be a random uncertainty ($s_{\bar{X}}$) if one were describing facility-to-facility variability. However, from the perspective of working in only one facility, that same data scatter indicates that one facility could have a systematic error compared to the others by an amount within the interval of $s_{\bar{X}}$ (the systematic uncertainty). The exact same error source is on one occasion random and on another systematic, depending on the context of the test.

Remember, error is the actual true difference between the measured data and the true value. One never knows the true value or the true error, but can only estimate its limits. This is called the uncertainty. In estimating the limits of a systematic error, use systematic standard uncertainty. For random error, use random uncertainty.

The following is the best rule to follow: *If an error source causes scatter in* the test result, it is a random error. All other error sources are systematic *errors.* This is without regard to the source of data used to estimate an uncertainty. The emphasis here is on the effect of the error source. It does not matter where the error comes from; it matters where it's going—its effect. What matters is the effect that the error source has on the manager's ability to make decisions with the test results. Errors have the effect of inhibiting proper decisions or, worse, selling a faulty product. Systematic errors, while invisible in test data, will cause an entire test program to be off from the true value. Random errors, their effects always visible in the test data, will cause scatter in the test results—always, by definition.

Categorizing error sources, therefore, is simply utilizing the above rule. It works in every case. It ignores origins. It considers what most experimenters want to know: What does an error or uncertainty do to the data or test result? This view allows one to apply the knowledge to the decision-making process that follows an experiment. After the above categorization, and after uncertainty estimates are made for the previously mentioned error sources, subscripts should be added to the elemental uncertainty values to denote the origin of the uncertainty estimates.

Designating Error Sources

A subscripted notation for error sources is described in Ref. 3. The general notation for an elemental systematic standard uncertainty is b_{ij}, where the b_{ij} are the systematic standard uncertainty estimate for the ith elemental error source in the jth category. Similarly, the notation for an elemental random error source (standard deviation) is s_{ij}, where s_{ij} is the standard deviation of the data from the ith error source in the jth category. Note that each s_{ij} would be divided by the appropriate N_{ij} to obtain the applicable random standard uncertainty, $s_{\overline{X},ij}$. These i and j subscripts are not utilized in this book to designate error sources in categories.

A simpler notation will work just fine. Note each uncertainty source with only one subscript to identify its source. Double subscripts may be used if a characterization into major groups is desired, but it is unnecessary in uncertainty analysis. Just denote each elemental systematic uncertainty as b_i and each random uncertainty as $s_{\overline{X},i}$. The single subscript method i will be utilized throughout this book to designate error or uncertainty sources.

3-3. The Calculation of Measurement Uncertainty (Symmetrical Systematic Uncertainties)

Combining Random Uncertainties

It is informative to note that usually there is no need to determine the effect of combined random errors once a test has been run. The scatter in the data is exactly the right effect of all the random error sources! The standard deviation of that scatter divided by the square root of the number of data points averaged is the random uncertainty, $(s_X/\sqrt{N})$. However, before scarce dollars are spent on a test program, it is important to know whether or not the results will be useful. In this important case, it is necessary to do a pretest uncertainty analysis as discussed in Unit 4. In this section, the methods for combining the effects of several uncertainty sources will be delineated. This combination is a requirement for understanding the expected uncertainty of a test before it is run.

The Standard Deviation: The Elemental Random Uncertainty

The first level of random uncertainty is the elemental random uncertainty. This is the standard deviation of the data caused by each random error source. The equation for the standard deviation of the elemental random error source is:

$$s_{X,i} = \left[\frac{\sum_{k=1}^{N_i} \left(X_{i,k} - \overline{X}_i\right)^2}{N_i - 1} \right]^{\frac{1}{2}} \tag{3-1}$$

Note: The sum is over k where there are N_i values of $X_{i,k}$.

Here, $s_{X,i}$ is the standard deviation of elemental random error source i. The standard deviation is calculated exactly as in Equation 3-1. If this were the only random error source, the test data would exhibit scatter identical to that described by $s_{X,i}$.

However, there is always more than one error source, and it is necessary to combine their effects when estimating what the random uncertainty for an experiment might be. Having combined them, it is necessary to estimate the appropriate degrees of freedom. The first step in this combination is the calculation of the standard deviation of the average, the random standard uncertainty, for an error source.

The Random Standard Uncertainty for an Error Source

What matters most about an error source is its average effect for a particular experimental result. Seldom does one care what the actual data scatter might be. One is always concerned about its average effect. In fact, test results are almost always reported as the average for the data obtained. The effect on the average for a particular elemental random uncertainty can be expressed as:

$$s_{\bar{X},i} = \frac{s_{X,i}}{\sqrt{N_i}} \tag{3-2}$$

where:

$\quad s_{\bar{X},i} =$ the *random standard uncertainty* (or standard error of the mean) for error source i

$\quad N_i \quad =$ the number of data points averaged for error source i

This is an outgrowth of the central limit theorem (Ref. 4). What is being done here is to estimate the standard deviation of the average from the standard deviation of the data. Note that N_i may or may not be the same as the N_i used to calculate the standard deviation. This denominator N_i is always the number of data points that are in the average, or test result.

What is most important is the effect that an elemental random error source has on the average value. Stated another way, what would the standard deviation of a group of averages be for a particular error source? Usually, there is neither time nor money to run numerous experiments so that many averages can be obtained. It is necessary, therefore, to estimate what that distribution of averages would be; in other words, what the standard deviation of that group of averages would be. That estimate is achieved with Equation 3-2, where the standard deviation of the data is used to estimate the standard deviation of the average for a particular elemental random error source.

The Standard Deviation of the Average for the Result: The Random Standard Uncertainty

It is then the combined effect of the several random uncertainties on the average for the test or the test result that needs evaluation. That combined effect is determined by root-sum-square as in Equation 3-3:

$$s_{\bar{X},R} = \left[\sum_{i=1}^{N_i} \left(s_{\bar{X},i}\right)^2 \right]^{\frac{1}{2}} \tag{3-3}$$

where $s_{\bar{X},R}$ and $s_{\bar{X},i}$ have the same units. Note that the sum is over i. It is the $s_{\bar{X},R}$ values that are root sum squared, not the $s_{X,i}$ values. This is an important fact to remember throughout the remainder of this book. We will always root-sum-square $s_{\bar{X}}$ (s-sub-X-bar) values.

s_R is sometimes called the standard deviation of the average for the result and it is the *random standard uncertainty for the result*. It forms the basis for the random part of the *uncertainty statement*. Note that $t_{95}\, s_R$ could be called the random uncertainty component of the uncertainty statement.

Combining Degrees of Freedom

In order to utilize the correct t_{95} (Student's t) in an uncertainty statement, it is necessary to know the correct degrees of freedom, v_R, associated with the *random standard uncertainty for the result*, $s_{\bar{X},R}$. Each of the elemental random standard uncertainties, $s_{\bar{X},i}$, has its associated degrees of freedom, v_i. The same degrees of freedom, v_i, are associated with the corresponding standard deviation of the average for the error source, $s_{\bar{X},i}$. When the $s_{\bar{X},i}$ are combined with Equation 3-3, it is necessary to combine degrees of freedom as well. Ref. 5 and 6 utilize the Welch-Satterthwaite approximation given originally in Ref. 7. There (in the absence of any systematic errors) the degrees of freedom for a test result are given as:

$$v_R = \frac{\left[\sum_{i=1}^{N} \left(s_{\bar{X},i}\right)^2 \right]^2}{\left[\dfrac{\left(s_{\bar{X},i}\right)^4}{v_i} \right]} \tag{3-4}$$

Remember:

$$s_{\overline{X},i} = \frac{s_{X,i}}{\sqrt{N_i}}$$

(3-5)

Remember too that:

$$s_{X,i} = \left[\frac{\sum_{i=1}^{N_i}\left(X_i - \overline{X}\right)^2}{N_i - 1}\right]^{\frac{1}{2}}$$

(3-6)

Combining Symmetrical Systematic Standard Uncertainties

Systematic standard uncertainties may be expressed as *symmetrical*, the first case treated here, or *nonsymmetrical*, which will be presented later.

Systematic standard uncertainties come from various error sources, and their combined effect must be evaluated. This is true whether the effect is to be noted in a pretest uncertainty analysis or a post test analysis. Remember, test data, while revealing the magnitude and effect of random errors, do not do so for systematic errors. One cannot see the effect of systematic errors in test data. It must be estimated from the combined effect of various sources.

The Symmetrical Systematic Standard Uncertainty for the Test Result

Each elemental systematic standard uncertainty, b_i, must have its combined effect on the test result evaluated. That combined effect is determined by Equation 3-7:

$$b_R = \left[\sum_{i=1}^{N}(b_i)^2\right]^{1/2}$$

(3-7)

where b_R is the systematic uncertainty of the test result and where b_R and b_i have the same units.

b_R is the basis for the systematic standard uncertainty component of the uncertainty statement. It is the value used for the symmetrical systematic standard uncertainty. Note here that the systematic standard uncertainties, whether elemental or for the result, represent one standard deviation of the average as do the random standard uncertainties.

Summarizing Uncertainty Categorization (Symmetrical Systematic Standard Uncertainties)

Systematic and random error sources and the estimates of their limits, systematic and random standard uncertainties, and their categories may be summarized as shown in Table 3-1. Assuming that the values in Table 3-1 are the uncertainties at the temperature calibration point of interest, the systematic standard uncertainty for the calibration result is obtained as follows:

$$b_R = \left[\sum_{i=1}^{3} (b_i)^2 \right]^{\frac{1}{2}} = \left[(0.025)^2 + (0.005)^2 + (0.01)^2 \right]^{\frac{1}{2}} = 0.027 \qquad (3\text{-}8)$$

To obtain the random standard uncertainty for this experiment, a similar rootsum-square approach is used as follows:

$$
\begin{aligned}
s_R &= [\Sigma(s_{X,i}/(N_i)^{1/2})^2]^{1/2} \\
&= [\Sigma(s_{\bar{X},i})^2]^{1/2} \\
&= [(0.056)^2 + (0.016)^2 + (0.045)^2]^{1/2} \\
&= 0.074°F
\end{aligned}
\qquad (3\text{-}9)
$$

Table 3-1. Temperature Calibration Uncertainties

Error Sources (i)	°F Systematic Standard Uncertainty (b_i)	°F Standard Deviation $(s_{X,i})$	No. Data Points Avg'd. (N_i)*	°F Random Standard Uncertainty $(s_{\bar{X},i})$	Degrees of Freedom $(N-1)$ (v_i)
(1) Intermediate thermocouple reference	0.025B	0.056A	1	0.056A	29
(2) Ice reference junction	0.005B	0.016A	1	0.016A	9
(3) Voltmeter readout	0.01B	0.1A	5	0.045A	4

*Number of data points averaged in this measurement. That is, 1 intermediate reference reading, 1 ice reference reading, and 5 voltmeter readings averaged for each temperature measurement.

Note: The degrees of freedom are associated with $s_{X,i}$ and $s_{\bar{X},i}$ and not N_i. Also, the subscripts A and B have been used to denote the existence or nonexistence of data to calculate the standard deviation for the uncertainty source shown. For all cases when an uncertainty is Type B, infinite degrees of freedom are assumed.

To compute the uncertainty, it is necessary to obtain a t_{95} value from Appendix D. To do that, the appropriate degrees of freedom must be obtained using the Welch-Satterthwaite approximation, Equation 3-4. Note that the systematic uncertainties, Type B, must also be included, with degrees of freedom assumed to be infinite. This is because, you will recall, each b_i represents a 68% confidence estimate of a normal distribution of errors. So, each b_i represents one standard deviation.

$$v_R = \frac{[\Sigma(s_{\bar{X},i})^2 + \Sigma(b_i)^2]^2}{\{\Sigma[(s_{\bar{X},i})^4/v_i] + [\Sigma(b_i)^4/v_i]\}}$$

$$v_R = \frac{[(0.056)^2 + (0.016)^2 + (0.045)^2 + (0.025)^2 + (0.005)^2 + (0.01)^2]^2}{[(0.056)^4/29] + [(0.016)^4/9] + [(0.045)^4/4] + 0 + 0 + 0}$$

$$= 27.7 = 27$$

The three zeros in the denominator are a result of the degrees of freedom for the b_i terms being infinite.

Note that, to obtain the correct degrees of freedom, 27.7 is truncated to 27. Always truncate, as this is the conservative value for the degrees of freedom. The evaluation of why this is conservative is left as an exercise for the reader (see Exercise 3-5).

Appendix D provides $t_{95} = 2.052$ for 27 degrees of freedom.

The information is now available to form an uncertainty statement.

3-4.   The Uncertainty Statement (Symmetrical Systematic Standard Uncertainties)

Measurement uncertainty is defined as the combination of both the systematic and random components of uncertainty, formerly called bias and precision. An appendix in Ref. 5 provides one method for that combination.

However, it is instructive to note two older methods that, while not now utilized, do appear in many books and papers. These older models are the "Addition Model" and the "Root-Sum-Square Model." They are presented here for information only and will not be used in this book.

The Addition (ADD) Uncertainty Model

The addition uncertainty model is defined as:

$$U_{ADD} = \pm[b_R + t_{95}s_{\bar{X}, R}] \tag{3-10}$$

where U_{ADD} is the additive uncertainty. Here B_R equals $2*b_R$.

U_{ADD} provides an interval (or coverage) around the test average that will contain the true value ~99% of the time (Ref. 8). It is also called U_{99} for that reason. Stated another way, the true value, μ, is contained in the interval:

$$(\overline{X} - U_{ADD}) \le \mu \le (\overline{X} + U_{ADD}) \tag{3-11}$$

approximately 99% of the time. This interval may also be written as:

$$(\overline{X} - U_{99}) \le \mu \le (\overline{X} + U_{99}) \tag{3-12}$$

Note:

$$U_{ADD} = U_{99} \tag{3-13}$$

The Root-Sum-Square (RSS) Uncertainty Model

The root-sum-square uncertainty model is defined as:

$$U_{RSS} = \pm\left[(B_R)^2 + (s_{\overline{X},R})^2\right]^{\frac{1}{2}} \tag{3-14}$$

where U_{RSS} is the root-sum-square uncertainty. Here B_R again equals $2*b_R$.

U_{RSS} provides an interval around the test average that will contain the true value ~95% of the time (Ref. 8). It is also called U_{95} for that reason. Stated another way, the true value, μ, is contained in the interval:

$$(\overline{X} - U_{RSS}) \le \mu \le (\overline{X} + U_{RSS}) \tag{3-15}$$

approximately 95% of the time.

However, Ref. 5 recommends newer methods, two improved uncertainty models: U_{95} and U_{ASME}. The origin and detailed development of these models are discussed in Refs. 1 and 2.

The U_{95} and U_{ASME} Uncertainty Models

In this book, the U_{ASME} model will be used in those cases where the degrees of freedom may be assumed to be 30 or higher. The U_{95} model will be used all other times. Note that U_{95} and U_{ASME} are both ASME models.

The U_{ASME} model is:

$$U_{ASME} = \pm 2.00[(b_R)^2 + (s_{\overline{X},R})^2]^{1/2} \tag{3-16}$$

This model provides 95% confidence under a vast variety of conditions (Ref. 9) and is the recommended model for easing uncertainty calculations. This is the case because the "2.00" in front of the equation is Student's t for 30+ degrees of freedom and 95% confidence. This simple model may also be altered to other confidences by using the U_{95} model:

$$U_{95} = \pm t_{95}[(b_R)^2 + (s_{\overline{X},R})^2]^{1/2} \tag{3-17}$$

Here there is no assumption of the degrees of freedom being 30 or higher, and an individual Student's t is calculated for each uncertainty statement. This model provides whatever confidence is desired by the experimenter. It does so by the proper selection of Student's t. That is, for 99% confidence and 30+ degrees of freedom, $t_{99} = 2.58$. For 99.7% confidence, $t_{99.7} = 3.0$. For any confidence and less than 30 degrees of freedom, a standard statistics text will contain the table from which Student's t can be obtained. The degrees of freedom for the U_{95} uncertainty model are obtained with the Welch-Satterthwaite approximation, in which the b_i are recognized as equivalent to one standard deviation of the average.

Although Equation 3-17 is a more robust uncertainty equation to use, the use of Equation 3-16 instead of 3-17 will seldom lead to difficulty in real world situations (Refs. 10, 11).

It should be noted that some utilize the phrases "total uncertainty," "expanded uncertainty" and "combined uncertainty." We will note only that "combined uncertainty" will be taken to mean the root-sum-square of the random standard uncertainty and the systematic standard uncertainty. The "total uncertainty" and "expanded uncertainty" will be considered synonymous and be taken to mean the "combined uncertainty" multiplied by the appropriate Student's t to obtain the 95% confidence interval.

Assumption of "b" Usage Throughout This Text

Note that in the above expressions for uncertainty, Equations 3-16 and 3-17, the systematic uncertainty is noted as a b term. This is because it is assumed that all the systematic standard uncertainties represent error sources that are normally distributed. In addition, the b term usage presumes that all the systematic standard uncertainties have infinite degrees of freedom, and are estimated at 68% confidence. This will be the assumption throughout this book unless otherwise noted.

NOTE NOW THAT THIS CONDITION FOR THE b TERMS ALTHOUGH *MOST OFTEN* THE CASE, WILL NOT *ALWAYS* BE THE CASE IN

PRACTICE. There are times when systematic uncertainties, estimated at 68%, will not posses infinite degrees of freedom. A specific example of this would be the confidence one has in a calibration curve fit. In that case, the 95% confidence interval on the curve fit is estimated as:

$$Conf_{95} \approx t_{95} * (SEE)/\sqrt{N}$$

Where N = the number of data points in the curve fit.

This $Conf_{95}$ is the systematic uncertainty of using the curve fit calculated line. When this term is incorporated into the calculation of the systematic uncertainty of the result, it should be entered as $Conf_{95}/t_{95}$. In this way, that term usually noted as b will maintain its equivalent $1s$ value for all root-sum square operations to follow.

However, that being said, we will assume throughout this text, that we will assume that people make systematic uncertainty estimates at 95% confidence for normally distributed error sources with infinite degrees of freedom. That is, in the absence of any data, estimates of systematic uncertainty are typically approximately 95% confidence. This is taken as typical of individuals' estimates when they are requested to make those estimates without data. Other confidence intervals could be chosen but it is recognized that the most common occurrence is 95% confidence.

The U_{ISO} Uncertainty Model

Note that the calculation of uncertainty using the strict ISO approach yields the same result. In the strict ISO approach we must evaluate U_A and U_B.

Using the example of results from Table 3-1, we note that $U_A = s_{\overline{X},R}$ and that $U_B = b_R$. It is b_R because all the b_i in Table 3-1 are 68% confidence estimates for a normal distribution and thus, an equivalent, $1s$ values.

Since $U_{ISO} = k[(U_A)^2 + (U_B)^2]^{1/2}$ there really is no difference in result between the U_{95} and the U_{ISO} where the ISO "k" multiplier is taken as Student's t. In addition, there is no difference between the U_{ASME} and the U_{ISO} where "k" is taken as "2."

3-5.   Computing the Uncertainty Interval (Symmetrical Systematic Uncertainties)

The information is now available to compute the uncertainty interval, or measurement uncertainty, for the case in Table 3-1 with symmetrical systematic uncertainties. The values to be used are from Equations 3-8 and 3-9.

The U_{95} Uncertainty Interval (Symmetrical Systematic Uncertainty)

For the U_{95} uncertainty, the expression is:

$$U_{95} = \pm 2.052[(b_R)^2 + (s_{\bar{X}, R})^2]^{1/2}$$
$$= \pm 2.052[(0.027)^2 + (0.073)^2]^{1/2}$$
$$= \pm 0.16°F$$

Assuming an average calibration correction of +2.00°F at the temperature of interest, it can be stated that the interval:

$$(2.00 - 0.16) \leq \mu°F \leq (2.00 + 0.16)$$
$$1.84 \leq \mu°F \leq 2.16$$
$$\mu = 2.00 \pm 0.16°F$$

contains the true value for the correction, μ, 95% of the time; that is, there is 95% confidence that the true value is contained in the above interval.

Note that the calibration constant, or correction, is not the uncertainty or the error. The uncertainty is now how well the calibration correction [+2.00F] is known [±0.016F].

The concept of confidence is shown schematically in Figure 3-1 for the U_{ASME} model (which is the easiest to show graphically). Of importance to note is that the uncertainty interval is widened by both the random and the systematic components of uncertainty. In Figure 3-2, the total (~68% conf. not expanded) uncertainty interval of 2.00 ± 0.16 is shown; in this interval, the true value of the temperature calibration correction will be contained 95% of the time.

The U_{ASME} Uncertainty Interval (Symmetrical Systematic Uncertainty)

For the U_{ASME} uncertainty, the expression is:

$$U_{ASME} = \pm 2[(b_R)^2 + (s_{\bar{X}, R})^2]^{1/2}$$
$$= \pm 2.00[(0.027)^2 + (0.073)^2]^{1/2}$$
$$= \pm 0.16°F$$

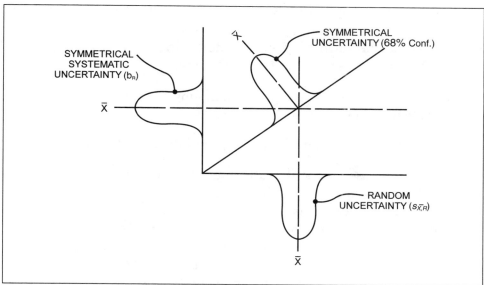

Figure 3-1. ~68% Conf. Symmetrical Uncertainty Interval

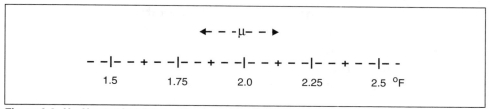

Figure 3-2. U_{95} Uncertainty Interval for $\mu = 2.00 \pm 0.16$

This is exactly the same as the U_{95} uncertainty. If there were fewer degrees of freedom, the resultant U_{95} uncertainty would be greater than the U_{ASME} because a larger Student's t would be needed.

The Choice of the Uncertainty Model

The uncertainty model should be chosen for the confidence desired. In some industries, such as aerospace, a great penalty is assessed if a wrong decision is made based on a test result. Therefore, in the past, the aerospace standard (Ref. 12) has utilized the U_{ADD} model, which provides 99% coverage of the true value. In the past, the steam turbine world, however, preferred the 95% coverage of the U_{RSS} model.

Now and in the future, the choice of Student's t and the U_{95} model will afford the analyst the confidence of interest and still allow the use of the most robust uncertainty model, that is, U_{95}. This is true for the aerospace applications and the steam turbine applications just cited.

The choice is up to the user, but the model chosen must be reported.

3-6.   The Calculation of Measurement Uncertainty (Nonsymmetrical Standard Systematic Uncertainties)

The first step here is the proper handling of the nonsymmetrical systematic standard uncertainties.

Combining Nonsymmetrical Systematic Standard Uncertainties

For most cases, the systematic standard uncertainties and the resulting uncertainty interval or measurement uncertainty will be symmetrical; that is, there is an equal risk of error in either the positive or the negative direction. However, there are times when the systematic standard uncertainty is not symmetrical. In these cases, something is known about the physics or engineering of the measurement that prompts a tendency for the errors to be either negative or positive.

An example of this kind of error might be a pressure reading from a transducer, which may tend to read low while trying to follow a ramp up in pressure. Another example might be a thermocouple in a process stream that, because of the laws of thermal conductivity, will always tend to read low when reading higher than ambient temperatures and high when reading lower than ambient temperatures.

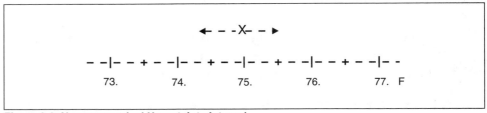

Figure 3-3. Nonsymmetrical Uncertainty Interval

Summarizing Uncertainty Categorization (Nonsymmetrical Systematic Standard Uncertainties)

Suppose that, of the uncertainties in Table 3-1, the intermediate thermocouple reference calibration systematic standard uncertainty was nonsymmetrical, in that the tendency was to read low for the higher than ambient calibration points. Given that the systematic standard uncertainty of $\pm 0.025°F$ was nonsymmetrical as follows: -0.04 to $+0.01°F$, it is noted that the min. to max. range of systematic standard uncertainty is still $0.05°F$. However, it is skewed negative. The uncertainty estimates then need to be reformatted from Table 3-1 into Table 3-2. This concept is shown schematically in Figure 3-4.

Table 3-2. Temperature Calibration Uncertainty Sources

Error Sources (i)	°F Systematic Standard Uncertainty (b_i^-)	°F Systematic Standard Uncertainty (b_i^+)	°F Standard Deviation $(s_{X,i})$	No. Data Points Avg'd. $(N_i)^*$	°F Random Standard Uncertainty $(s_{\bar{X}, i})$	Degrees of Freedom (N − 1) (v_i)
(1) Intermediate TC reference	-0.04	+0.01	0.056	1	0.056	29
(2) Ice reference junction	0.005	0.005	0.016	1	0.016	9
(3) Voltmeter readout	0.01	0.01	0.1	5	0.045	4
*See note on N_i following Table 3-1.						

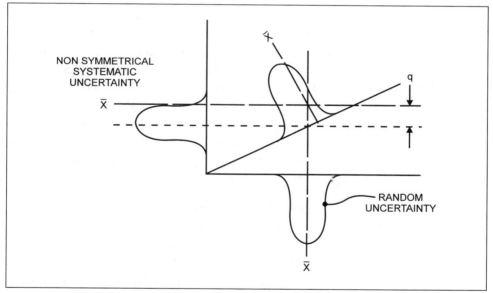

Figure 3-4. Uncertainty Interval for 74.24°F ≤ $\bar{X}$ ≤ 75.36°F

In Table 3-2 note the use of the notation b_i^- and b_i^+. The superscripts minus and plus indicate the negative and positive systematic uncertainties.

Since its systematic standard uncertainty is symmetrical, the ice reference junction systematic standard uncertainty is reported as ±0.1°F. When there are some elemental systematic standard uncertainties that are nonsymmetrical, the proper handling of the summations requires that all systematic standard uncertainties be defined as minus and plus, or upper and lower limits. The ice reference junction systematic standard uncertainties, therefore, need to be noted as $b^- = 0.1°F$ and $b^+ = 0.1°F$, as shown in Table 3-2.

Systematic and random error sources and their uncertainties may be summarized as shown in Table 3-2.

Assuming the uncertainties noted are the uncertainties at the temperature calibration point of interest, the systematic standard uncertainties for the calibration are obtained as follows. Recalling Equation 3-7:

$$b_R = \left[\sum_{i=1}^{N} (b_i)^2 \right]^{1/2} \tag{3-7}$$

The nonsymmetrical systematic standard uncertainty is computed using the methods of Ref. 13. This approach will be in the new U.S. national standard (Ref. 2). It proceeds as follows: Consider that for a normally distributed uncertainty source, its mean error may be represented as a displacement, q, from the central value such that a symmetrical systematic standard uncertainty interval may then be estimated about the displacement, $\overline{X} - q$. The symmetrical systematic standard uncertainty thus obtained (I love that expression) is then root-sum-squared with all the other symmetrical systematic standard uncertainties, a symmetrical total uncertainty interval is calculated, and then the displacement, q, is used to reinsert the nonsymmetry so the resulting final total uncertainty interval is nonsymmetrical. (Follow all that?)

If there is more than one nonsymmetrical elemental systematic standard uncertainty, several q values will be obtained and their use will create several symmetrical systematic standard uncertainties to root-sum-square for the final symmetrical systematic standard uncertainty. The several q values are then added algebraically and used to offset the resulting symmetrical total uncertainty the proper amount to make it nonsymmetrical.

To see how this works, consider the following equations and the example of the uncertainties in Table 3-2. (Translation: These are the steps to follow for one nonsymmetrical systematic standard uncertainty source.)

First, the generalized equations are needed:

1. Given $s_{\overline{X}}$, b^+, and b^- (the nonsymmetrical systematic standard uncertainty) and the 68% confidence systematic uncertainty interval: $(\overline{X} - b^-)$ to $(\overline{X} + b^+)$

2. Let q equal the difference between the center of the nonsymmetrical systematic uncertainty interval and the average. That is:

$$q = [(\overline{X} + b^+) + (\overline{X} - b^-)]/2 - \overline{X} \qquad (3\text{-}18)$$

3. Estimate a symmetrical b as one-half the nonsymmetrical systematic standard uncertainty interval:

$$b = [(\overline{X} + b^+) - (\overline{X} - b^-)]/2 \qquad (3\text{-}19)$$

4. Complete the total uncertainty estimate for 30+ degrees of freedom:

$$U_{95} = \pm 2.00[(b_R)^2 + (s_{\overline{X},R})^2]^{1/2}$$

Recall that b_R and $s_{\overline{X},R}$ represent the root-sum-squares of the individual b's and $s_{\overline{X}}$'s. In this nonsymmetrical case, the b from Equation 3-19 is one of the b's root-sum-squared to obtain b_R.

5. Compute the 95% confidence uncertainty interval, I_{95}:

$$I_{95} = (\overline{X} + q) \pm U_{95} \qquad (3\text{-}20)$$

6. Compute the nonsymmetrical, 95% confidence, total uncertainty interval, I, about the average:

$$I_{lower} = \overline{X} - (U_{95} - 2q) = \overline{X} - U^- \qquad (3\text{-}21)$$

where I_{lower} is the lower limit of the nonsymmetrical 95% confidence total uncertainty interval.

$$I_{upper} = \overline{X} + (U_{95} + 2q) = \overline{X} + U^+ \qquad (3\text{-}22)$$

where I_{upper} is the upper limit of the nonsymmetrical 95% confidence total uncertainty interval.

Note that:

$$U^- = U_{95} - q \tag{3-23}$$

and that

$$U^+ = U_{95} + q \tag{3-24}$$

Example 3-1:

To see how this works, consider the data in Table 3-2. Going through the six steps shown above, we have the following:

1. For the first uncertainty source, the one with nonsymmetrical systematic uncertainty, $s_{\overline{X},1} = 0.056°F$, $b_1^+ = 0.01°F$, and $b_1^- = 0.04°F$. Here assume the average for the test or experiment is 75°F, $\overline{X} = 75°F$.

2. $q_1 = [(75 + 0.01) + (75 - 0.04)]/2 - 75 = -0.015$ (from Equation 3-18)

    Note the subscript 1 on q to denote the q from uncertainty source 1. Although there is only one nonsymmetrical systematic uncertainty in this example, were there more than one, the q would later need to be summed algebraically.

3. $b_1 = [(75 + 0.01) - (75 - 0.04)]/2 = 0.025$ (from Equation 3-19)

4. $U_{95} = \pm 2.00[(b_R)^2 + (s_{\overline{X},R})^2]^{1/2}$ \hfill (3-16)

    $b_R = [\Sigma(b_i)^2]^{1/2} = \pm[(0.025)^2 + (0.005)^2 + (0.001)^2]^{1/2} = 0.026°F$

    $s_{\overline{X},R} = [\Sigma(s_{\overline{X},i})^2]^{1/2} = \pm[(0.056)^2 + (0.016)^2 + (0.045)^2]^{1/2} = 0.074°F$

    Therefore, assuming 30+ degrees of freedom:

    $U_{95} = \pm 2.00[(0.026)2 + (0.074)2]1/2 = \pm 0.16°F$

5. Compute the 95% confidence interval, I_{95} (from Equation 3-20):

    $I_{95} = [(75 + (-0.03)) \pm 0.16]°F$

6. The nonsymmetrical total uncertainty interval is then (from Equations 3-21 and 3-22):
    $I_{lower} = 75 - [0.16 - (-0.03)] = 74.87°F$
    $I_{upper} = 75 + [0.16 + (-0.03)] = 75.13°F$

Here, $U^- = 0.16 - (-0.015) = 0.175°F$ from Equation 3-23 and $U^+ = 0.16 + (-0.015) = 0.145°F$ (from Equation 3-24).

Therefore, the average and its uncertainty interval would be written as: $(75 - 0.175)°F$ to $(75 + 0.145)°F$, or the 95% confidence uncertainty interval is 74.825 to 75.145°F.

Figures 3-3 and 3-4 illustrate this nonsymmetrical uncertainty interval.

This is clearly a nonsymmetrical uncertainty interval about the average, 75°F. It is a 95% confidence interval.

3-7.   Common Uncertainty Model Summary

A common uncertainty model is summarized in Table 3-3 for the case of three categories and three uncertainty sources in each category. Note that here there are double subscripts under each uncertainty source, i and j, to designate an uncertainty source, i, in a category, j. No mention of Type A or B is shown here for clarity.

Table 3-3 shows clearly the summation and handling of the degrees of freedom. It can be used as a guide for as many categories as needed, with as many uncertainty sources as possible in each category, if grouping uncertainty sources in categories is desired. Remember, this is not necessary at all and some feel it is an unnecessary complication of the uncertainty estimation process. Table 3-3 is the uncertainty calculation for a single measurement or the measurement of a single parameter, such as temperature.

The b_R and $s_{\bar{X},R}$ are then used in the uncertainty calculation as appropriate.

If the result is not a simple parameter measurement such as temperature or pressure, Table 3-3 is used for each parameter measurement used to calculate a result; for example, b_R, s_R, and v would be calculated for temperature, pressure, etc., and those values propagated into the final result uncertainty. This uncertainty propagation is covered in Unit 5.

If only the uncertainty of a single measurement result (e.g., temperature) is needed, the uncertainty is as shown in Table 3-3. To go further and determine the uncertainty for several results, the random uncertainty would again be divided by the square root of the number of results, M; that is, for multiple results:

$s_{\overline{X}, R}/\sqrt{M}$ = the random error component of the uncertainty for several, M, results for one parameter such as temperature or pressure

M = the number of results to be averaged for a parameter

Table 3-3. Uncertainty Summary for Three Uncertainty Categories with Three Elemental Uncertainties Sources Each

Category	Systematic	Random	Degrees of Freedom
1	$\left.\begin{array}{l}b_{11}\\b_{21}\\b_{31}\end{array}\right\}$ RSS $= b_1$	$\left.\begin{array}{l}s_{X,11}/\sqrt{N_{11}}=s_{\overline{x},11}\\s_{X,21}/\sqrt{N_{21}}=s_{\overline{x},21}\\s_{X,31}/\sqrt{N_{31}}=s_{\overline{x},31}\end{array}\right\}$ RSS $= s_{\overline{x},1}$	$\left.\begin{array}{l}\nu_{11}\\\nu_{21}\\\nu_{31}\end{array}\right\}$ W/S $= \nu_1$
2	$\left.\begin{array}{l}b_{12}\\b_{22}\\b_{32}\end{array}\right\}$ RSS $= b_2$	$\left.\begin{array}{l}s_{X,12}/\sqrt{N_{12}}=s_{\overline{x},12}\\s_{X,22}/\sqrt{N_{22}}=s_{\overline{x},22}\\s_{X,32}/\sqrt{N_{32}}=s_{\overline{x},32}\end{array}\right\}$ RSS $= s_{\overline{x},2}$	$\left.\begin{array}{l}\nu_{12}\\\nu_{22}\\\nu_{32}\end{array}\right\}$ W/S $= \nu_2$
3	$\left.\begin{array}{l}b_{13}\\b_{23}\\b_{33}\end{array}\right\}$ RSS $= b_3$	$\left.\begin{array}{l}s_{X,13}/\sqrt{N_{13}}=s_{\overline{x},13}\\s_{X,23}/\sqrt{N_{23}}=s_{\overline{x},23}\\s_{X,33}/\sqrt{N_{33}}=s_{\overline{x},33}\end{array}\right\}$ RSS $= s_{\overline{x},3}$	$\left.\begin{array}{l}\nu_{13}\\\nu_{23}\\\nu_{33}\end{array}\right\}$ W/S $= \nu_3$
	$\downarrow$ RSS $= b_R$, the systematic uncertainty	$\downarrow$ RSS $= s_{\overline{x},R}$, the random uncertainty	$\downarrow$ W/S $= \nu$, the degrees of freedom of $s_{\overline{x},R}$

Note: Only RSS errors with the same units. Propagate errors into result units before combining with RSS.
W/S = the Welch-Satterswaithe method for combining degrees of freedom. As with RSS, all the $s_{\overline{x}}$'s must have the same units.
RSS = the root-sum-square

Uncertainty Category	Systematic Uncertainty	Random Uncertainty	Degrees of Freedom	~Random Uncertainty of Source
1	b_1	$s_{\overline{x},1}$	ν_1	$t_{95}\, s_{\overline{x},1}$
2	b_2	$s_{\overline{x},2}$	ν_2	$t_{95}\, s_{\overline{x},2}$
3	b_3	$s_{\overline{x},3}$	ν_3	$t_{95}\, s_{\overline{x},3}$
	$\downarrow$ RSS $= b_R$	$\downarrow$ RSS $= s_{\overline{x},R}$	$\downarrow$ W/S $= \nu_R$	

Note: b_R = the systematic uncertainty of the result
$s_{\overline{x},R}$ = the random uncertainty of the result
ν_R = the degrees of freedom of the uncertainty, say, U_{95}
t_{95} = the Student's t for degrees of freedom for $s_{\overline{x},R}$

3-8.   Summary

This unit has presented the basics of the error and elemental uncertainty categorization process and uncertainty statement. It is now possible for the student to calculate the measurement uncertainty for simple measurement systems.

References

1. *Guide to the Expression of Uncertainty in Measurement*, 1993. Geneva, Switzerland: International Standards Organization.

2. ANSI/ASME PTC 19.1-2006, *Performance Test Code, Instruments and Apparatus, Part 1, Test Uncertainty.*

3. ANSI/ASME PTC 19.1-2006, *Performance Test Code, Instruments and Apparatus, Part 1, Measurement Uncertainty*, p. 15.

4. Bowker, A. H., and Lieberman, E. H., 1969. *Experimental Statistics*, pp. 99–101. Englewood Cliffs, NY: Prentice-Hall.

5. ANSI/ASME PTC 19.1-1998, p. 20.

6. Coleman, H. W., and Steele, W. G. Jr., 1989. *Experimentation and Uncertainty Analysis for Engineers*, p. 97. New York: John Wiley & Sons.

7. Brownlee, K. A., 1967. *Statistical Theory and Methodology in Science and Engineering*, 2nd ed., p. 300. New York: John Wiley & Sons.

8. Abernethy, R. B., and Ringhiser, B., 1985. "The History and Statistical Development of the New ASME-SAE-AIAA-ISO Measurement Uncertainty Methodology." Presented at AIAA/SAE/ASME/ASEE 21st Joint Propulsion Conference, Monterey, CA.

9. Steele, W. G.; Ferguson, R. A.; Taylor, R. P.; and Coleman, H. W., 1994. "Comparison of ANSI/ASME and ISO Models for Calculation of Uncertainty." *ISA Transactions* 33:339–352.

10. Strike, W. T. III, and Dieck, R. H., 1995. "Rocket Impulse Uncertainty; An Uncertainty Model Comparison." In *Proceedings of the 41st International Instrumentation Symposium*, Denver, CO. Research Triangle Park, NC: ISA.

11. Dieck, R. H., 1996. "Measurement Uncertainty Models." In *Proceedings of the 42nd International Instrumentation Symposium*, San Diego, CA. Research Triangle Park, NC.

12. Abernethy, R. B., et al., 1973. *Handbook—Gas Turbine Measurement Uncertainty*, AEDCTR-73-5, Arnold AFB, TN: Arnold Engineering Development Center.

13. Steele, W. G.; James, C. A.; Taylor, R. P.; Maciejewski, P. K.; and Coleman, H. W., 1994. "Considering Asymmetric Systematic Uncertainties in the Determination of Experimental Uncertainty." AIAA Paper 94-2585, presented at the 18th AIAA Aerospace Ground Testing Conference, Colorado Springs, CO.

Exercises:

3-1.   Systematic Uncertainty Problems

Consider the following list of systematic standard uncertainties: 1, 2, 3, 10.

a.   What would you get after combining them to obtain the systematic standard uncertainty of the test result?

   1) 10.68   2) 106.00   3) 13.74   4) _____

b.   Same as (a), but with systematic standard uncertainties of: 1, 2, 3, 10, 11, 12.

   1) 19.47   2) 39.00   3) 36.74   4) _____

c.   What is learned about the impact of large uncertainties?

3-2.   Random Uncertainty Problems

Consider the following random uncertainties for two uncertainty sources:

$$s_{\overline{X},1} = 1.0\% \qquad v_1 = 10 \text{ degrees of freedom}$$

$$s_{\overline{X},2} = 2.0\% \qquad v_2 = 20 \text{ degrees of freedom}$$

a.   What is the combined random uncertainty, $s_{\overline{X},R} = (s_{\overline{X},1}^2 + s_{\overline{X},2}^2)^{1/2}$?

b.   What are the equivalent degrees of freedom? (Note: Use the Welch-Satterthwaite approximation.)

c.   What is $t_{95} * s_{\overline{X},R}$? (This is the random uncertainty component of the result.)

3-3.   Combining Degrees of Freedom Problem

Write the formula for the standard deviation of the result, $s_{\overline{X},R}$, that combines the individual random uncertainties, $s_{\overline{X},i}$.

3-4.   Truncating Degrees of Freedom Problem

Why is truncating the fractional degrees of freedom obtained with the Welch-Satterthwaite approximation conservative?

3-5.   Nonsymmetrical Uncertainty Interval Problem

Given $s_{\overline{X}} = 4$, $b^+ = 1.5$, $b^- = 3$ and $\overline{X} = 26$, calculate the U_{95} (95% confidence) nonsymmetrical n that s uncertainty interval (the upper and lower nonsymmetrical systematic uncertainties).

3-6.   Scale and Truth Problem

Consider every possible uncertainty source for a measurement process that consists of daily weighings on your bathroom scale in preparation for your annual physical examination when you will be weighed on your doctor's scale. Categorize the uncertainty sources into systematic and random.

a. What is truth in this process?

b. How would you calibrate your scale?

c. How would you estimate the random uncertainty?

d. Make your best estimate of the systematic and random uncertainties of a typical bathroom scale.

Unit 4:
How To Do It Summary

UNIT 4

How To Do It Summary

In this unit, nomenclature will be explained, specific steps used in an uncertainty analysis will be emphasized, and details will be presented on how to estimate both systematic and random uncertainties. Step-by-step instructions along with the treatment of calibration uncertainties and the timing of an uncertainty analysis will also be considered. This unit will reinforce the fundamentals of the statistics and data analysis needed to complete a credible measurement uncertainty analysis.

Learning Objectives—When you have completed this unit you should:

A. Understand and be able to compute systematic and random standard uncertainties.

B. Be able to properly combine elemental systematic and random standard uncertainties.

C. Understand the proper treatment of calibration uncertainties.

D. Understand the need for and be able to complete both a pretest and a posttest uncertainty analysis.

E. Understand the statistical basis for uncertainty analysis.

While some of the methods presented in this unit are covered in parts of other units, it is sometimes easier for the student to have a central reference for this material. Hence, this information is presented in a convenient format.

4-1. Review of Nomenclature

The nomenclature used thus far has covered s_X, $s_{\overline{X}}$, X, $\overline{X}$, N, b, t_{95}, and U. These are the generic parameters whose definitions can be summarized as follows:

$$
\begin{aligned}
s_X &= \text{the standard deviation of a data set or sample} \\
s_{\overline{X}} &= \text{the standard deviation of the sample average, the random} \\
&\quad \text{standard uncertainty} \\
X &= \text{the individual data point, sometimes called } X_i \\
\overline{X} &= \text{the average of the data set or sample of } X_i\text{'s} \\
N &= \text{the number of data points used to calculate } s_X
\end{aligned}
$$

N = the number of data points in the average $\overline{X}$, sometime redefined after the calculation of s_X

b = the systematic standard uncertainty component of the uncertainty

t_{95} = the Student's t statistic at 95% confidence

U = the uncertainty of an experiment or measurement

All the above are nomenclature and symbols used to describe a data set or sample and error in a data set. Except for N and t_{95}, none of them are exact. They all are estimates of the true value of the population and are obtained from the sample of data taken.

The population true values are:

σ_X = the standard deviation of the population

$\sigma_{\overline{X}}$ = the standard deviation of a set of averages taken from the population

ε = the actual random error of a data point; it is a value never known and is estimated with sX, the random uncertainty

β = the true systematic error for a data point or set of data; it is a value never known and is estimated with b, the systematic standard uncertainty

δ = $\varepsilon + \beta$, the true total error of a data point; it is a value never known and is estimated by U, the measurement uncertainty

The above true values are never known but can only be estimated, using the variables mentioned. This is summarized in Table 4-1. As can be seen, an experimenter or measurement specialist never knows the true values for errors. Only estimates are known, as drawn from the sample of data taken.

Table 4-1. Comparison of Population and Sample Statistics [The true error compared with the estimate of the true error, the uncertainty.]

Error Type	True Error	Uncertainty
Random	ε	$s_X/\sqrt{N}$ or $s_{\overline{X}}$
Systematic	β	b
Total	$\delta = \varepsilon + \beta$	U

Therefore, the material that follows concentrates on sample statistics and the statistical basis for the calculation of uncertainty.

4-2.   How To Do It Summary

Now that the reader has been introduced to the basics, this section will
outline the techniques used to obtain an uncertainty analysis. There are
seven parts to this approach:

A. Calculation of random standard uncertainties

B. Obtaining and combining systematic standard uncertainties

C. Step-by-step instructions

D. Treatment of calibration uncertainties

E. Making pretest estimates of uncertainty and why

F. Making posttest estimates of uncertainty and why

4-3.   Calculation of Random Standard Uncertainties

This section presents several ways to obtain the random standard
uncertainty. All are based on the standard deviation of the data, s_X.

$$s_X = \left[\frac{\sum_{i=1}^{N}(X_i - \bar{X})^2}{N-1} \right]^{\frac{1}{2}} \qquad (4\text{-}1)$$

The standard deviation of the data is calculated with Equation 4-1. It is not
the random uncertainty, which is calculated using Equation 3-2.
Generically, that equation is:

$$s_{\bar{X}} = s_X / \sqrt{N} \qquad (4\text{-}2)$$

Remember, the random uncertainty is the estimate of the standard
deviation of the average. It is an estimate of the standard deviation of a
group of averages from samples of data with the same number of data
points as the sample taken to yield the $s_{\bar{X}}$ already obtained from the data.

The key then is to first obtain the standard deviation of the data. There are
at least four ways to do that:

A. When the variable being measured can be held constant and "N"
    repeat measurements are taken

B. When there are "M" redundant measurements, each with "N" repeat measurements

C. When there is a pair of redundant measurements of the same variable at the same time

D. When there is applicable calibration data

Calculating s_X When the Variable Being Measured Can Be Held Constant

When the measured variable can be held constant, the standard deviation of the data can be used to calculate the random uncertainty of the measurement system by utilizing Equation 4-2. This is the simplest and the most common case. After a test, for example, the variability in the data may be expressed as s_X. The random standard uncertainty follows immediately. The random uncertainty is the expected scatter, were redundant averages possible to obtain.

Calculating s_X When There Are "M" Redundant Measurements, Each with "N" Repeats

This is a special case of a more common method called *pooling*. The general expression for s_X when there are M redundant measurements with N repeats is:

$$s_X = \left[\frac{\sum_{i=1}^{N} \sum_{k=1}^{M} \left(X_{i,k} - \overline{X}_k \right)^2}{((MN)) - M)} \right]^{\frac{1}{2}} \tag{4-3}$$

Here, the double sum is first over k to M and then over i to N. Equation 4-3 is actually a special case when all M groups of data have the same number of data points, N. The general formula for pooling is:

$$s_{X, pooled} = \left[\frac{\sum_{i=1}^{N} v_i \left(s_{X,i} \right)^2}{\sum_{i=1}^{N} v_i} \right]^{\frac{1}{2}} \tag{4-4}$$

Pooling data from several (N) sets or samples, as is done in Equation 4-4, provides a better estimate of the standard deviation of the population of data than any one sample can provide. Note the degrees of freedom for the pooled standard deviation is just the sum of all the degrees of freedom, or, the denominator in Equation 4-4. It is recommended that Equation 4-4 be utilized all the time. It never fails.

Again, once the standard deviation of the data is obtained, the random uncertainty follows easily. Just divide the resulting s_X by the square root of the number of data points averaged when obtaining $\overline{X}$. Note again, that the N used here is related to the number of data points averaged, not necessarily the number of data points or the degrees of freedom in the various samples.

The degrees of freedom for Equation 4-4 constitute the sum of the degrees of freedom for all the N samples. It is the defining parameter used to obtain t_{95} in an uncertainty analysis. It is needed to obtain the random standard uncertainty, $t_{95}\, s_{\overline{X}}$.

Calculating s_X When There Is a Pair of Redundant Measurements of the Same Variable at the Same Time

It is often not realized that instrument or measurement system variability may be estimated with no knowledge of the measurement system or variable being measured except for: (1) the assumption that two identical instruments or measurement systems have the same random uncertainty and (2) the assumption that the variable being measured is the same for both instruments or systems. For example, if it is desired to estimate the random uncertainty of two CO_2 analyzers, it can be done without knowing the level of CO_2 being measured. All that is needed is to assume that the instruments respond identically and that both instruments see the same process stream.

In this case, the test data are analyzed to determine the standard deviation of the deltas between the instruments, s_Δ, which is defined by Equation 4-5:

$$s_\Delta = \left[\frac{\sum_{i=1}^{N} (\Delta_i - \overline{\Delta})^2}{(N-1)} \right]^{\frac{1}{2}} \tag{4-5}$$

Equation 4-5 is just the standard deviation of a set of data where all the i data points are deltas.

It is important to note then that s_Δ is related to the standard deviation of either the instruments or the measurement systems. They are related by Equation 4-6:

$$s_X = s_\Delta / \sqrt{2} \tag{4-6}$$

The proof for this equation will be an exercise for the reader in Unit 5.

It can be seen, therefore, that as long as an instrument and/or a measurement system sees the same variable at the same time and as long as they have the same variability, the standard deviation for the data from either may be estimated with Equations 4-5 and 4-6.

The random standard uncertainty, $s_{\overline{X}}$, is then obtained using Equation 4-7:

$$s_{\overline{X}} = s_X / \sqrt{N} \tag{4-7}$$
$$= s_\Delta / (\sqrt{2N})$$

Equation 4-7 will prove to be a valuable tool when simultaneous measurements from two instruments are available. It is used after a test is run and the delta data are obtained. N is the number of test data points that will be averaged into a test result. Yes, the random standard uncertainty depends not only on the standard deviation of the data but also on the number of data points to be averaged.

Here the degrees of freedom associated with $s_{\overline{X}}$ is one less than the number of data points used to calculate s_Δ.

Calculating s_X When There Is Applicable Calibration Data

Another method for estimating the random standard uncertainty of an instrument is to use calibration data. This is effective only when the calibration process may be deemed to yield instrument performance identical to that obtained on the process stream.

That one constraint on instrument performance can be ascertained from calibration data taken in the field or on the test site itself with a portable standard. This will work as long as the standard has a variability that is significantly lower than the test instrument. In this way all the variability evident in the on-site calibration can be attributed to the measurement instrument or system and not to the standard.

In this case, the random uncertainty is obtained simply from the standard deviation of the calibration data. It is important to realize here that it is assumed that only one level is being calibrated in this discussion. For the proper treatment of more than one level or calibration constant, called a calibration curve or line, see Section 7-4.

Once the standard deviation of the calibration data is obtained, one merely divides it by the square root of N, the number of test data points that will be averaged into a test result. This is shown in Equation 4-8:

$$s_{\overline{X}} = s_{CAL}/\sqrt{N} \qquad (4\text{-}8)$$

Remember here that s_{CAL} is the standard deviation of the calibration data for a single calibration constant. N is the number of test data points that will be averaged into the test result. Here, too, the degrees of freedom for $s_{\overline{X}}$ are the same as that for s_{CAL}.

4-4.   Obtaining and Combining Systematic Standard Uncertainties

Obtaining Systematic Standard Uncertainties

It is possible that the most difficult part of any uncertainty analysis is obtaining systematic standard uncertainties. Frequently there are no data from which to calculate a systematic standard uncertainty.

Ref. 1 recommends five possible ways to obtain systematic standard uncertainties (called the antiquated term "bias limits" in Ref. 1):

  A. Interlaboratory comparisons

  B. Standards in the test environment

  C. Auxiliary or concomitant functions

  D. Systematic uncertainties from special causes, special tests

  E. Judgment or estimation

The above are in rough order of preference to the experimenter. The first is by far the best as it has a solid basis in data. The last is by far the worst, relying on judgment alone with all its foibles.

Interlaboratory Comparisons

Almost all experimenters or metrologists agree that the best way to get a handle on their own facility's systematic uncertainties is to run interfacility comparisons or tests. These tests, often called round robins, require a stable set of standards or measurement processes that can be sent from one place to another without affecting its average value or variability. Each facility in turn then measures this artifact and the results from all the facilities are analyzed, usually by an independent analyst; it is also usually the case that the analyst does not know which test result is from which participating facility or laboratory.

The results from this kind of series of tests may be taken two ways. First, the variability observed between laboratories, $s_{X,IL}$ (standard deviation interlaboratory), may be used to estimate the variability across laboratories. That is, approximately 95% of the results from various laboratories will be contained in the interval $\overline{X} = t_{95}s_{X,IL}$. It is then used as an estimate of the systematic uncertainty for any one laboratory, $t_{95}s_{X,IL}$. Note that $s_{X,IL}$ is not divided by the square root of N laboratories, because it is for only one laboratory that we are interested in the possible variability from the average for all the laboratories. If we were interested in the expected variability of the average of all the laboratories, we'd use $s_{X,IL}/\sqrt{N}$.

Standards in the Test Environment

Sometimes it is possible to take a mobile standard out to the test environment and calibrate the measurement system in place. This method of calibration is similar to the Measurement Assurance Program (MAP) of the National Institute of Standards and Technology (NIST).

In this method of systematic uncertainty estimation, a carefully documented, stable set of calibration artifacts is sent to the calibration laboratory for calibration. There is a significant difference from a normal calibration, however. In the normal case, an instrument or measurement system is calibrated by comparing it to a standard, usually much more accurate than the instrument under calibration, in order to determine a calibration constant or calibration curve for the instrument. In this application, the standard is assumed to be more accurate than the instrument, and the correction or calibration curve is then used to correct test data to be more closely aligned with the standard.

The calibration process is assumed to have very small or no systematic errors, and the standard deviation of the data may be used to assess the random uncertainty. However, it is known that every measurement process has systematic error. The standard instrument may be calibrated, but the process in which it is used in the laboratory cannot be calibrated by normal means. This is the strength of the MAP. The carefully monitored portable standards are sent to the calibration laboratory and calibrated as though they were test instruments. The big difference is that the errors observed are attributed to the calibration process and the laboratory standard, not to the "instrument" or calibration artifact being calibrated.

By calibrating a standard in the laboratory, the errors in the process itself may be identified and quantified. Such is the case in any test when a standard may be used to calibrate the instruments or measurement systems at the test site. The errors observed may be used to assess the systematic uncertainty of the measurement process itself. The standard is assumed correct, except for its documented systematic uncertainty. The process itself—instrument, standards, and methods—are presumed to account for the remaining differences observed. The differences observed can be used to estimate the systematic uncertainty of the measurement process.

Auxiliary or Concomitant Functions

It is sometimes the case that systematic error for several independent methods can be estimated by comparing the test results obtained when using all the methods on the same process stream. It is important that the measurement methods be independent; that is, their error sources must have no relationship from one method to another. These independent methods are sometimes called concomitant methods.

As an example, air flow in a gas turbine test may be determined by an orifice in the inlet duct, an inlet bell mouth (nozzle) on the engine, a compressor speed-flow map, the turbine flow parameter, and the exhaust nozzle coefficient. Each of these measurements may be set up to be independent. The degree of disagreement between them may be used to estimate the systematic uncertainty typical of any. That is, assuming each has the same systematic uncertainty, the scatter between the air flow measured may be used to estimate a single systematic uncertainty assignable to any of them.

Example 4-1:

It may be assumed that the values for gas turbine air flows obtained by the five methods might be as follows:

Measurement Method	Value Obtained
Pipe orifice	230.1 lb/s
Inlet bell mouth	224.7 lb/s
Compressor map	245.8 lb/s
Turbine flow parameter	232.4 lb/s
Nozzle coefficient	228.7 lb/s

For the above data, $\overline{X} = 232.3$ and $s_X = 8.0$. The systematic standard uncertainty for each of the concomitant methods. (Removing known random error as well is considered in Section 7-7.)

Systematic Uncertainty from Special Causes, Special Tests

Special tests can always be designed to evaluate the systematic standard uncertainties that may exist in a measurement process. Special calibrations may be run to perturb the parameter of interest throughout its range, thus revealing potential systematic error sources and estimates of the systematic standard uncertainty. Usually, the experimenter will be the best person to suggest ways to evaluate what the systematic uncertainty might be.

Judgment or Estimation

By far the worst method for estimating the magnitude of a systematic standard uncertainty is judgment. When there are no data or the process is so unique that special calibrations or tests are not available, the judgment of the engineer, experimenter, or researcher must be used.

Here, the problem is pride. Many experimenters will not admit that their experiment or measurement system has any likelihood of any error, let alone the possibility that significant systematic error occurs that would displace all their measurements from the true value. Often the experimenter will point to the low data scatter or low random uncertainty and assume there is no systematic error to speak of. This must be countered with the argument that some systematic error always exists and must be estimated, with the systematic standard uncertainty.

It should also be noted that the truly competent experimenter does have a good idea of the magnitude of the measurement system errors. What is needed is to get him or her to reveal true feelings about their magnitude.

Throughout, however, this remains the worst method for systematic standard uncertainty estimation. It often yields unreasonably small uncertainties and always yields systematic standard uncertainties that are difficult to defend to a third party.

Avoid this method at all costs!

Combining Systematic Uncertainties

There are three major methods in the current literature for combining systematic standard uncertainties. Only one is recommended and has legitimacy from an established standard. These three methods follow, with the least acceptable first and the recommended one last.

Summing Systematic Uncertainties

The simplest method for combining systematic standard uncertainties is to simply sum them. Equation 4-9 illustrates the mathematics of this simplest approach:

$$b_R = [\Sigma(b_i)] \qquad (4\text{-}9)$$

The sum is over the number of error sources, i.

This method is heartily not recommended, as it yields a systematic standard uncertainty much too large to be reasonable. Consider, for example, the following three elemental systematic uncertainties:

$$b_1 = 5.0_A°F$$
$$b_2 = 7.3_B°F$$
$$b_3 = 4.5_B°F$$

Note that the subscripts A and B have been used to conform to the ISO approach, which identifies the source of the uncertainty estimate as either having data to estimate a standard deviation, A, or not, B.

Using Equation 4-9, the systematic standard uncertainty would be:

$$b_{sum} = (5.0 + 7.3 + 4.5)°F = 16.8°F$$

This limit has meaning. If twice the systematic standard uncertainties represent 95% confidence estimates of the limits systematic error (d.f. = 30^+), then having an estimate of their combined effect of 33.6°F means that the individual elemental errors have high probability of each being at their extreme limit, or upper 95/lower 5 percentile. For the first elemental systematic uncertainty, there is 1 chance in 20 that it will be at one or the other extreme. There is 1 chance in 40 that it will be high and one chance in 40 that it will be low. The same is true for the second and third. For all three to be at their same either upper or lower extremes, the condition modeled by summing the systematic uncertainties, the odds are $(1/40)^3$, or 1 in 64,000! This is extremely unlikely and much poorer odds than the recommended 95% confidence systematic uncertainty confidence of about 1 in 20 chances of the true systematic error being outside the quoted symmetrical 95% confidence systematic uncertainty limit.

The use of this method is obviously unreasonable.

Summing Major Elemental Systematic Standard Uncertainties and Root-Sum-Squaring Minor Ones

In the second method, which involves summing major elemental systematic uncertainties and root-sum-squaring minor ones, the square root is taken of the square of the sum of the three or four largest elemental systematic standard uncertainties combined with the sum of the squares of the rest. Equation 4-10 expresses this model:

$$b = [\{\Sigma(\{\text{largest}\}b_i)\}^2 + \{\Sigma(b_i)^2\}]^{1/2} \qquad (4\text{-}10)$$

This is also an unreasonable approach, which can be understood by examining Example 4-2.

Example 4-2:

Consider the three elemental systematic standard uncertainties presented previously and add to them the following three:

$$b_4 = 0.2_B°F$$
$$b_5 = 0.1_B°F$$
$$b_6 = 0.3_A°F$$

Again, note the subscripts A and B for ISO harmony.

With this approach, the first three elemental systematic uncertainties would be summed to yield $(b_{sum}) = 16.8°F$ as before; the sum of the squares of the last three would yield (b_{RSS}) as follows:

$$(b_{RSS})^2 = (0.2^2 + 0.1^2 + 0.3^2) = 0.14°F$$

Combining these two terms into Equation 4-10, the following is obtained:

$$b = (16.8^2 + 0.14)^{1/2} = 16.8°F$$

This result is as unreasonable as the first model's. The additional terms obtained from the three small errors are lost completely in the impact of the sum of the largest three. This, too, is an extremely unlikely result for a systematic standard uncertainty.

Root-Sum-Square of the Elemental Systematic Standard Uncertainties

The recommended method is to root-sum-square the elemental systematic uncertainties. Its calculation is expressed by Equation 3-7:

$$b = [\Sigma(b_i)^2]^{1/2}$$

Its application was illustrated in the previous unit but will be reviewed here.

Consider the first three elemental systematic uncertainties above. For them,

$$b = (5.0^2 + 7.3^2 + 4.5^2)^{1/2} = 9.9°F$$

This result is reasonable as it allows for the distinct possibility that some elemental systematic errors (their limits are merely expressions of their possible range) will be positive and others negative, thus having the effect of partial cancellation; hence, this result, which is more than the largest systematic standard uncertainty (expected) but much less than the linear sum, which is so unlikely. Since all these elemental systematic standard uncertainties are 68% confidence, twice them is 95% confidence, or one chance in 20 that the actual systematic error is greater than the plus or minus twice the systematic standard uncertainty shown (d.f. = 30[+]).

Were the calculation to include the small systematic standard uncertainties, they would have little effect even in this model. This is left as an exercise for the student.

Summary

The recommended combination method for systematic standard uncertainties is the root-sum-square approach, as recommended by the ASME standard (Ref. 2).

4-5.   Nonsymmetrical Systematic Standard Uncertainties

In the preceding discussions, the assumption was made that the systematic standard uncertainties were symmetrical. That is, they were ± the same absolute amount. Although this is not always the case, the recommended approach to their combination does not change. What is affected is the impact on the uncertainty band for the systematic error portion of the uncertainty, the systematic standard uncertainty.

As is the case of a symmetrical systematic standard uncertainty, the true value for the measurement or experiment likely lies within the distance from the average and the average plus or minus the systematic standard uncertainty. However, what is missed in this argument is that the true value is in the range of the average plus the positive systematic standard uncertainty and the range of the average minus the negative systematic standard uncertainty. A negative systematic standard uncertainty means the measurement is below the true value. A positive systematic standard uncertainty means the measurement is above the true value.

This distinction is completely lost for symmetrical standard systematic uncertainties and no one cares. However, for nonsymmetrical systematic standard uncertainties, the effect can be significant. It is easier to understand when the random errors are zero and will be explained as such here for clarity and reinforcement.

Suppose the final systematic standard uncertainty is nonsymmetrical and is expressed as:

$$b^- = 5.0°F$$
$$b^+ = 2.0°F$$

In the case where there is no random error, the true value is in the range of $(\overline{X} - |b^-|)$ up to $(\overline{X} - |b^+|)$, or $(\overline{X} - 5.0°F)$ up to $(\overline{X} + 2.0°F)$ with 68% confidence.

Remember that each b represents a 68% confidence estimate, so b approximates one standard deviation.

As shown in Unit 3, nonsymmetrical systematic standard uncertainties lead to nonsymmetrical uncertainty intervals.

4-6.   Step-by-Step Instructions

Although this ILM is not intended to be a cookbook for uncertainty analysis, some specific steps should be taken when embarking on such an analysis. These generalized steps are outlined in Ref. 3 and described below.

Step 1. Define the measurement system.

The purpose of this step is to consider all aspects of the test situation: the calibrations, the duration, the instruments, the data acquisition, the error (or uncertainty) propagation (covered in Unit 5), the data reduction, and the presentation of results (covered in Unit 8). The test objective should be clearly stated, including the success criteria—that is, what level of what parameter must be achieved to declare the test or experiment a success. *Do not allow yourself to just run the test or* experiment. Consider the results and then decide upon the success or failure of *the new process.* Do this honestly and do it before the test.

Include in that assessment of success criteria the acceptable level of uncertainty. Go into an experiment or test with a result goal that includes a maximum permissible measurement uncertainty. Note that it may be necessary here to carefully define the performance or calculation equations to do this right.

This process, then, is usually done before a test. It is only then that this approach is needed to define the random uncertainties. After a test is run, the data themselves will usually be sufficiently voluminous to permit a direct calculation of the standard deviation of the result average (random uncertainty) from the test results. This process should still be applied, even after testing, to obtain the systematic uncertainty.

Step by step, this part of the process includes the following:

    A.   List all the independent parameters and their expected levels.

    B.   List all the required calibrations and how the data are to be reduced.

    C.   List and define all the data reduction calculations required to obtain the test result from the measurement parameters.

Step 2. List all the error sources for which you'll make uncertainty estimates.

The intent is to compile a complete, exhaustive list of all possible error sources, systematic and random, without regard to their possible size or impact on the experimental result. Here the intent is also to include in the data the subscripts A or B to designate whether or not data are available for an uncertainty or error source to calculate a standard deviation. This is the ISO harmonization process. Step by step, this part of the process is as follows:

A. List all the uncertainty sources.

B. If convenient, group the uncertainty sources into calibration, data acquisition, data reduction, and errors of method categories. This step is not a necessity for a viable uncertainty analysis, but it often makes the analysis itself easier and clearer.

C. Further group the error sources into systematic and random standard uncertainties. Here, a simple rule will help: *If the error source will cause scatter in your test result, it is a random error source whose limits are estimated by the random standard uncertainty; if not, it is a systematic error source whose limits are estimated by a systematic standard uncertainty.* This rule will never fail and should be used with confidence.

Step 3. Estimate the magnitude of all the elemental uncertainty sources.

Here the random standard uncertainties may be estimated from previous test data or inferred from calibration data. The systematic standard uncertainties should be obtained in one of the five suggested ways discussed earlier in this unit under "Obtaining Systematic Standard Uncertainties."

Step 4. Calculate the systematic standard uncertainties and random standard uncertainties for each parameter in the test.

Combine systematic and random standard uncertainties to obtain the systematic standard uncertainty and the random standard uncertainty for each parameter measured in the experiment or test. The methods of Unit 3 should be used for this combination of uncertainties.

Step 5. Propagate the systematic standard uncertainty and the random standard uncertainty (standard deviation of the average) into the test result using the methods of Unit 5.

It is important to remember to ascertain whether the errors being combined are independent or not. The methods of Unit 3 are only for independent error sources and, happily, most measurement parameters are. However, when the same error source shows up in several places in a test result, the error sources are not independent. The propagation formulas that are designed for that situation must be used. Those also are given in Unit 5. Note here that it is the dependency of error sources considered in Unit 5, although it is the uncertainties that are propagated. More on that later.

Step 6. Calculate the uncertainty.

The second-to-last step is to calculate the uncertainty. This is merely the square root of the sum of $(b)^2$ and $(s_{\bar{X}})^2$, the result multiplied by 2 for the U_{ASME} model.

Step 7. Report the uncertainty.

When reporting the test or experimental result uncertainty, it is important also to independently report the systematic uncertainty, the random uncertainty, and the degrees of freedom that are associated with U_{95}. It is important to also report the uncertainty model used; that is, U_{95} or U_{ASME}. Unit 8 provides a detailed treatment of the reporting of measurement uncertainty.

4-7.   Treatment of Calibration Errors (Uncertainties)

When considering the steps to an uncertainty analysis, it is important to carefully examine the calibration uncertainties. Single calibrations, multiple calibrations, and back-to-back tests all utilize calibration data, the same calibration data, differently.

Single Calibration, Short Test

In this application, the test or experiment is run over a time interval that is short. In this case, all the calibration errors manifest themselves only once, in the single calibration constant or curve. This constant or curve never changes for the duration of the experiment. This is the definition of a short test: The calibration constant or curve never changes.

Here, despite the fact that both systematic and random errors contribute to the uncertainty of the calibration, all the error of the calibration results in a systematic error for the rest of the experiment. Not even the random errors in the calibration process contribute to scatter in the test result.

It may be thought of this way: If the calibration constant or curve never changes for the experiment, whatever error has displaced it from its true value also does not change for the duration of the experiment. It is a constant, or systematic, error. The actual magnitude of this error is not known—only its limits, the uncertainty. The proper expression of the limits of this kind of calibration error is called fossilized uncertainty; that is, all the error of the calibration process is fossilized into systematic standard uncertainty for the test or experiment.

The equation to be used for the case of many degrees of freedom is:

$$b_{CAL} = [(b)^2 + (s_{\bar{X}})^2]^{1/2} \tag{4-11}$$

Here, all the error of the calibration process has been fossilized into a systematic standard uncertainty term for the rest of the calibration hierarchy. Equation 4-11 is in generic form. The actual systematic and random standard uncertainties so utilized are the combination of all the elemental uncertainty sources in the calibration process.

Note that this fossilization process often should occur more than once. It occurs at every level in the measurement hierarchy where a constant, average, or calibration curve is calculated and remains unchanged for the rest of the measurement hierarchy below it. However, the number of fossilizations that occur does not affect the final total uncertainty of the measurement process when Equation 4-11 is used.

Many Calibrations, Long Test

When there is a long test with many calibrations, the random errors in the calibration process will contribute scatter in the long-term test results. This is because there will be many calibrations throughout the test process. In this case, the usual root-sum-square of elemental systematic and random standard uncertainties is done as shown in Unit 3 for the calibration uncertainty sources along with the other uncertainty sources.

Here the calibration uncertainties are combined into the systematic and random portions of the uncertainty right along with the other uncertainty sources. The applicable equations are:

$$(b_{CAL}) = [\Sigma(b_i)^2]^{1/2} \tag{4-12}$$

$$s_{\bar{X},CAL} = \{\Sigma(s_{X,i})^2\}^{1/2} \tag{4-13}$$

In both Equations 4-12 and 4-13, only calibration uncertainties are considered.

Zero Calibration Errors, Back-to-Back Tests

One very effective way to see clearly a small effect in test results is to employ the technique called back-to-back testing. In this case, a test or experiment is run; then, rapidly, a change in the test object is made and the test is repeated. The time here must be short, and no recalibration of the measurement system between tests is allowed.

The test must be short enough so that it can be assumed that no instrument or measurement system drift or decalibration occurs. In addition, no recalibration of the instruments or measurement system may occur or the calibration constant or curve will shift. This will violate a fundamental principle of back-to-back testing; that is, the systematic errors of the calibration process do not have time or opportunity to shift between the two tests. If this is true, and the test time is short, all the calibration error is systematic as explained above. In addition, *all the calibration error is zero!* This rather surprising result occurs because the same systematic error that affects the first test also affects the second. Since back-to-back tests are designed to measure differences, not absolute levels, any systematic error in the calibration will not affect the difference and is zero by definition.

If the first test is high by 10 psi, the second test will be also, and the difference in psi will be unaffected by the systematic error from calibrations.

This technique should be employed whenever small changes need to be observed in large parameters.

4-8.   Simplified Uncertainty Analysis for Calibrations

Providing a detailed analysis of a calibration process is often fraught with difficulties related to harmonizing the analysis with the engineering uncertainty models of the ASME and the ISO. Decisions as to what systematic uncertainties and what random uncertainties are also either ISO Type A or Type B can unnecessarily complicate the process. This Section provides the simplifying assumptions needed to speed the uncertainty analysis for most calibrations. The resulting uncertainties will apply to the calibration results and is in full agreement with current Standards (Ref. 1) and Guidelines (Ref. 5 and 6).

The calculation methods in this Section are intended to apply to all calibrations. Three methods are given: one for calibrations which yield only the conclusion that the instrument under calibration is "in spec," one for calibrations that produce a calibration constant or curve and for which there are calibration data that are used in the uncertainty calculation, and, a third for which uncertainty propagation and degrees of freedom calculations are required. In the first two cases, all the units of the test data and uncertainty estimates are identical, that is for example, all degrees C, psia, m/sec, etc. In the third case, calibration units are not those of the final measurement and must be converted by some relationship or, uncertainty propagation.

The first is called the "In-Spec Calibration Uncertainty Calculation," the second the "Basic Calibration Uncertainty Calculation," and the third the "Propagated Calibration Uncertainty Calculation."

Assumptions Applicable to all Methods

Although each method has its own assumptions, the following assumptions are applicable to all three methods and greatly simplify the calculation of calibration uncertainty.

1. All degrees of freedom are 30 or higher.

2. All uncertainties are symmetrical about the measured average.

3. All systematic standard uncertainties are equivalent to ISO "Type B" uncertainties, are assumed to arise from normally distributed errors and are quoted at 68% confidence.

4. All random standard uncertainties are equivalent to ISO "Type A" uncertainties (except as otherwise noted) and are assumed to be one standard deviation of the average for a process or a measurement.

5. The U_{ASME} model is the appropriate choice.

In-Spec Calibration Uncertainty Calculation

The following procedure may be used to compute the measurement uncertainty for an instrument whose calibration does not result in a calibration constant or curve but only results in assessing whether or not the instrument is "in spec." It will yield an "in-spec" uncertainty statement in agreement with ASME PTC19.1 Standard and the ISO and NIST Guidelines.

Some additional assumptions are appropriate here.

1. The uncertainty in the standard is the first systematic standard uncertainty source.

2. The manufacturer's quoted instrument uncertainty or accuracy specification for an instrument under calibration is assumed to be "Type B" and is further assumed to have X fraction systematic and $(1-X)$ fraction random standard uncertainty.

3. All uncertainties are expressed in the units of the output of the instrument under calibration.

4. All effects of hysteresis, non-linearity, drift, etc. are contained within the manufacturer's quoted instrument uncertainty or accuracy.

In-Spec Uncertainty Calculation

The measurement uncertainty for an instrument calibrated and determined only to be "within spec" is calculated as follows:

$$U_{ASME} = \pm 2[(b)^2 - (s_{\overline{X}_I})^2]^{1/2} \tag{4-14}$$

where

U_{ASME} = the 95% confidence uncertainty
b = the systematic standard uncertainty of the instrument under calibration.

$$b = [(b_{SD})^2 + (b_I)^2]^{1/2} \tag{4-15}$$

where

b_{SD} = the uncertainty of the standard and

$$b_I = [(X)^{1/2}U_I] \tag{4-16}$$

where

U_I = the manufacturer's quoted instrument uncertainty or accuracy specification for the instrument under calibration. It is assumed to be a 95% confidence estimate for a distribution that is normal.

$$U_I = 2\left[(b_I)^2 + (s_{\overline{X},I})^2\right]^{\frac{1}{2}} \tag{4-17}$$

$s_{\overline{X},I}$ = the random standard uncertainty of the instrument under calibration

$$s_{\overline{X},I} = [(1 - X)^{1/2}(U_I/2)] \tag{4-18}$$

Note that here, X, is the fraction of the manufacturer's uncertainty that may be considered systematic. The remaining fraction, $(1 - X)$, is the random component of the manufacturer's uncertainty

Basic Calibration Uncertainty Calculation

The following procedure may be used to compute the measurement uncertainty for an instrument whose calibration results in a calibration constant or curve. It will yield a "basic" instrument uncertainty statement in agreement with ASME PTC19.1 Standard and the ISO and NIST Guidelines.

Some more assumptions are proper here. They are:

1. The uncertainty of the standard is the systematic standard uncertainty of the instrument under calibration and there are no additional systematic uncertainties.

2. All uncertainties are expressed in the units of the output of the instrument under calibration.

The simplified uncertainty for any instrument calibration is calculated as follows:

$$U_{ASME} = \pm 2\left[(b)^2 + \left(s_X / \sqrt{N}\right)\right]^{\frac{1}{2}} \qquad (4\text{-}19)$$

where

$U_{ASME} =$ the 95% confidence uncertainty

$b \ =$ the uncertainty of the standard (at 68% confidence). It is the systematic standard uncertainty of the instrument under calibration.

$s_X \ =$ the standard deviation of the calibration data. (Here estimate it with the Standard Estimate of Error (*SEE*) if the calibration produces a line fit.)

$N \ =$ the number of data points in the average calibration constant or the number of data points in the calibration line fit.

To obtain an estimate for "b," utilize one of the following approaches. They are listed in order of recommended preference, the first being the most preferred.

1. The U_{95} uncertainty estimate for the calibration of the standard. This is obtained from previous calibrations of that standard.

2. The manufacturer's statement of instrument accuracy or uncertainty. Assume it is a 95% confidence estimate unless informed otherwise.

3.   An engineering estimate of the 95% confidence systematic uncertainty of the calibration process.

The standard deviation, s_X, for the instrument being calibrated is calculated as follows:

For an average calibration constant

$$s_X = \left[\frac{\sum\limits_{i=1}^{N} (X_i - \overline{X})^2}{N-1} \right]^{1/2}$$ (4-20)

where

X_i = the ith data point used to calculate the calibration constant
$\overline{X}$ = the average of the calibration data (the calibration constant)
N = the number of data points used to calculate s_X
$N-1$ = the degrees of freedom for s_X

For a calibration line fit

$$s_X \approx SEE = \left[\frac{\sum\limits_{i=1}^{N} (Y_i - Y_{iC})^2}{N-K} \right]^{1/2}$$ (4-21)

Note that Equation 4-21 is an approximation that will work just fine for uncertainty analysis.

where

Y_i = the ith data point in the calibration corresponding to X_i
Y_{ic} = the Y_i value of the curve fit corresponding to X_i
N = the number of data points used to calculate SEE
K = the number of curve fit coefficients
$N-K$ = the degrees of freedom for the SEE and s_X

It should be clearly noted that the above does not address linearity, hysteresis, multiple calibrations, etc. It is a simplified approach that could be used as a first estimate of calibration uncertainty.

Propagated Calibration Uncertainty Calculation

Sorry, but there are more additional assumptions and they are below.

1. Systematic Standard Uncertainty: Systematic Standard Uncertainty is the estimate of the expected limits of systematic error. Systematic errors are those errors that do not change for the entire calibration. Each elemental systematic standard uncertainty (b_i) is estimated as the 68% confidence interval of a normal distribution for the ith elemental systematic uncertainty source. Each b_i therefore represents an equivalent "$1s_X$" for that systematic error source. The degrees of freedom for each bi are assumed to be infinite.

2. Random Standard Uncertainty: Random Standard Uncertainty is the estimate of the expected limits of random error. Random errors are those errors which change during a calibration and have their origin in a population of possible errors that is Normally distributed. Random Standard Uncertainty $[s_X/(N)^{1/2}]$ is estimated as the standard deviation of the calibration data (s_X) divided by the square root of the number of points used for the average calibration constant $(N^{1/2})$. Where a calibration curve is determined, the Standard Estimate of Error (SEE) is used to estimate s_X.

Random error sources produce errors which add scatter to the calibration data. If an error source does not add scatter to the calibration data, it is a systematic error source.

Propagated Uncertainty Calculation

The calculation of the calibration uncertainty when uncertainty propagation is needed is done as follows.

For propagated systematic standard uncertainty estimation, proceed as follows:

1. The systematic standard uncertainty (formerly called "bias") is obtained for each elemental systematic error source. The elemental systematic standard uncertainty is noted as b_i for the ith source, and is assumed to represent normally distributed errors, quoted at 68% confidence.

2. Each b_i must be expressed in the units of the output of the instrument under calibration either directly or after being multiplied by a sensitivity, or influence coefficient, Θ_i. Θ_i is the ratio of the output of the instrument used as the standard to the output of the instrument under calibration. It is the partial

differential of the output of the instrument under calibration (cal_i) with respect to the output of the standard (std_i), or, $\partial\,(cal_i)/\partial\,(std_i)$. Then, $b_i = \Theta_i\, b_{std,i}$ where $b_{std,i}$ is in the units of the standard instrument and b_i in units of the instrument under calibration. In calibration, it is almost always the case that Θ_i is unity. This greatly simplifies the job of uncertainty estimation.

6.2.1.3 The systematic standard uncertainty of the instrument under calibration, b, is the root sum square of all the elemental systematic uncertainties. That is:

$$b = \left[\sum_{i=1}^{N} (\Theta_i b_i)^2 \right]^{1/2} \quad \text{or} \tag{4-22}$$

$$b_R = \left[\sum_{T=A}^{B} \left(\sum_{i=1}^{N} (\theta_{i,T} b_{i,T})^2 \right) \right]^{\frac{1}{2}} \tag{4-23}$$

(Although both expressions yield the same uncertainty when combined in the total uncertainty equations (Equations 4-25 or 4-27), Equation 4-23 above is more easily seen to be in harmony with ISO methods.)

3. b is the systematic uncertainty of the instrument under calibration and is a 68% confidence interval, or equivalent $1s_X$. ($1s_X$ is also an equivalent as there is only one sample of each systematic standard uncertainty in a test or experiment, that is, $N = 1$.) b is an equivalent $1s_X$.

4. The degrees of freedom for b are assumed to be infinite.

5. Estimates of elemental systematic standard uncertainties for the instruments used in the calibration process may be obtained from previous uncertainty analyses of those instruments, manufacturer's uncertainty statements (see 4.1.6) or, as a last resort, estimates made by the engineer responsible for the calibration.

6. In addition, if desired, the elemental systematic standard uncertainties, b_i, may carry a second subscript of either "A" or "B" to designate an uncertainty source that either has data to calculate a standard deviation or does not. That is, $b_{i,A}$ or $b_{i,B}$ may be used to denote systematic uncertainty sources that either have data to calculate a standard deviation, Type "A," or don't, Type

"B." It is possible then to root sum square by Type "A" or "B" groupings. This is illustrated in Equation 4-23.

For the estimation of random standard uncertainties, proceed as follows.

Random standard uncertainty may be obtained three ways:

1. For a calibration constant: the random standard uncertainty is obtained from N replicate measurements of the instrument under calibration and is calculated as the standard deviation of the data divided by the square root of the number of data points. It is written as:

$$s_{\overline{X}} = s_X / \sqrt{N} = \left[\frac{\sum\limits_{i=1}^{N} (X_i - \overline{X})^2}{(N-1)} \right]^{1/2} / \sqrt{N} \qquad (4\text{-}24)$$

where:

$s_{\overline{X}}$ = the random uncertainty for the calibration in units of the instrument under calibration
$\overline{X}$ = the average X_i (the calibration coefficient or constant)
X_i = the ith data point, and,
N = the number of data points in the calibration
$N-1$ = the degrees of freedom for s_X and $s_{\overline{X}}$

2. For a calibration curve, substitute the Standard Estimate of Error (SEE) for the standard deviation in Equation 1 above and divide by the square root of the number of calibration points in the curve.

Then obtain:

$$s_{\overline{X}} \cong SEE / \sqrt{N} = \left[\frac{\sum\limits_{i=1}^{N} (Y_i - Y_{c,i})^2}{N-K} \right]^{1/2} / \sqrt{N} \qquad (4\text{-}25)$$

where

Y_i = the ith value of Y corresponding to X_i
$Y_{c,i}$ = the Y_i calculated from the curve fit evaluated at each X_i
K = the number of constants (coefficients) calculated for the curve fit

$$N = \text{the number of data points in the curve fit}$$
$$N - K = \text{the degrees of freedom for the curve fit}$$

3. $s_{\bar{X}}$ may be calculated from historical data if necessary. In that case, Equation 12 becomes:

$$s_{\bar{X}} = s_X / \sqrt{M} = \left[\frac{\sum\limits_{i=1}^{N} (X_i - \bar{X})^2}{N-1} \right]^{1/2} / \sqrt{M} \qquad (4\text{-}26)$$

where

$$M = \text{the number of data points in the average used as the calibration constant or coefficient}$$

$$s_{\bar{X}} = \text{the random uncertainty for the calibration in units of the instrument under calibration}$$

$$\bar{X} = \text{the average } X_i \text{ (the calibration constant or coefficient)}$$

$$X_i = \text{the } i\text{th historical data point}$$

$$N = \text{the number of historical calibration data points used to calculate } s_X$$

$$N - 1 = \text{the degrees of freedom for } s_X \text{ and } s_{\bar{X}}$$

6.2.2.4 Throughout 6.2.2.1, 6.2.2.2 & 6.2.2.3 above, it has been assumed that there was data to calculate the standard deviations. Thus, for ISO format, the standard deviations and standard deviations of the average could be noted as: $s_{X,A}$ and respectively.

If there was no data, they should be noted as: $s_{X,B}$ and $s_{\bar{X},B}$ respectively.

Simplified Measurement Uncertainty Calculation:

First we must deal with "degrees of freedom." In the simplest case is shown below.

Throughout this simplified analysis, the degrees of freedom are assumed to be large, that is, ≥30.

Measurement Uncertainty Calculation: The simplified equation for measurement uncertainty at 95% confidence is:

$$U_{ASME} = \pm 2[(b)^2 + (s_{\bar{X}})^2]^{1/2} \qquad (4\text{-}27)$$

Here the "2" multiplier out front is Student's t_{95} for 95% confidence. If a different confidence is desired, just choose the appropriate Student's t_{xx}, e.g., for 99% confidence, use $t_{99} = 2.58$.

General case for degrees of freedom less than 30:

In the Generalized Measurement Uncertainty Calculation, the degrees of freedom are assumed to be <30. Here, it is necessary to calculate the degrees of freedom associated with U_{95} and use a table of Student's t. The Student's t table should be entered at the appropriate confidence for the uncertainty calculation.

The degrees of freedom are calculated with the Welch-Satterthwaite approximation as follows:

$$(d.f.)_R = V_R = \frac{\left[\left(\sum_{i=1}^{N}(\theta_i b_i)^2\right) + \left(\sum_{j=1}^{M}(\theta_j s_{\bar{X},j})^2\right)\right]^2}{\left[\frac{\sum_{i=1}^{N}(\theta_i b_i)^4}{\sum_{i=1}^{N}V_i} + \frac{\sum_{j=1}^{M}(\theta_j s_{\bar{X},j})^4}{\sum_{j=1}^{M}V_j}\right]} \tag{4-28}$$

where
$(d.f.)_{R}V_{R}$ = the degrees of freedom for the calibration result uncertainty
V_i = the degrees of freedom of $s_{\bar{X},i}$ (i.e., $M - K$) for a curve fit or $(M - 1)$ for a single coefficient or constant)
Θ = the sensitivity or influence coefficients

The degrees of freedom of b are assumed to be infinite throughout.

Now we address the uncertainty calculation. The equation used to calculate the measurement uncertainty for 95% confidence is then:

$$U_{95} = \pm t_{95}\left[(b_R)^2 + (s_{\bar{X},R})^2\right]^{\frac{1}{2}} = \pm t_{95}\left[\left(\sum_{i=1}^{N}(\theta_i b_i)^2\right) + \left(\sum_{j=1}^{M}(\theta_j s_{\bar{X},j})^2\right)\right]^{\frac{1}{2}} \tag{4-29}$$

Where t_{95} is a function of v and found in a Student's t table. Note the subscript, R, stands for "result." For alternative confidences, just look up the appropriate Student's t_{xx}.

The above uncertainty was using the US U_{95} model. It is also acceptable to utilize the ISO uncertainty model. When that is done, the uncertainty calculation is done as follows:

> The equation to use for Type "A" and Type "B" measurement uncertainty calculation is identical to that of Equation 4-29 above except that the root sum square operations are done for all Type "A" uncertainties and then for all Type "B" uncertainties. The resulting Type "A" uncertainty and Type "B' uncertainty are then root sum squared and multiplied by the Student's t for the confidence of choice to obtain the total uncertainty. (ISO calls this the "Expanded Uncertainty.") That equation for 95% confidence would then be:

$$U_{95} = \pm t_{95}\left[\sum_{i=1}^{N_A} (b_{i,A})^2 \sum_{i=1}^{N_B} (b_{i,B})^2 + (s_{\overline{X},A})^2\right]^{1/2} \qquad (4\text{-}30)$$

where

$N_A =$ the number of $b_{i,A}$ systematic uncertainties
$N_B =$ the number of $b_{i,B}$ systematic uncertainties
$t_{95} =$ the degrees of freedom for U_{95} as determined by Equation 4-31.

The degrees of freedom is again calculated with the Welch-Satterthwaite approximation as follows:

$$d.f. = \upsilon = \cfrac{\left[\sum_{i=1}^{N_A} (b_{i,A})^2 + \sum_{i=1}^{N_B} (b_{i,B})^2 + (s_{\overline{X},A})^2\right]^2}{\left[\cfrac{\left(\sum_{i=1}^{N_A} (b_{i,A})^2\right)^2}{\infty} + \cfrac{\left(\sum_{i=1}^{N_B} (b_{i,B})^2\right)^2}{\infty} + \cfrac{(s_{\overline{X},A})^4}{\upsilon_{s_{\overline{X},A}}}\right]} \qquad (4\text{-}31)$$

Equation 4-30 yields the exact same total uncertainty as does Equation 4-29.

4-9.   Business Decisions and the Impact of Measurement Uncertainty

Uncertainty Analysis, the Cornerstone of Correct Business Decisions

Uncertainty analysis is the cornerstone upon which the whole structure of decisions based on data is built. Data either used or supplied to customers that has no attached data quality metric is useless in research, development, custody transfer and other scientific or business endeavors. Understanding the quality of data secured and provided is fundamental to the proper usage of the information derived from that data. This section discusses the impact of poor data quality on business decisions based on test or evaluation data.

Everyone making a decision based on test or evaluation data performs an uncertainty analysis on that data. If the results are pleasing to the evaluator, he or she presumes the errors in that data are small compared to the results reported. If the results are displeasing to the evaluator, there is often an immediate presumption that the data are flawed, or, have large embedded errors. Only when an objective assessment of the data quality is done can one be assured that the data are useful and the results applicable to a decision process. These decisions directly affect the profitability of the company making them.

The two data quality assessment, or measurement uncertainty analysis models now receiving worldwide acceptance are that of the International Standards Organization, *Guide to the Expression of Uncertainty in Measurement*, ISO, and that of the U.S. National Standard, the American Society of Mechanical Engineers (ASME) *Performance Test Code (PTC) 19.1,Test Uncertainty*. Both these models recognize that a clear distinction must be drawn between the term "error" and the term "uncertainty." Both models define "error" as the difference between the measured value (or measurand) and the true value (or value of the measurand). Both models define "uncertainty" as an estimate of the limits to which an error can go with some confidence. The ISO model recommends no specific confidence level. The ASME model recommends a 95% confidence level for most situations. Either of these two models may be used to assess data quality, its measurement uncertainty, and apply that assessment to the decisions business must make.

Impact of Data Quality, Measurement Uncertainty, on Business Decisions:

Although engineers and scientists prefer to understand the uncertainty of their test data, its data quality, management needs to apply this knowledge to business decisions. Without a clear understanding of measurement uncertainty and its impact at the management and executive levels of a corporation, business decisions based on test and evaluation data may be erroneous and very costly.

As an example, understanding the quality of custody transfer data in the gas and oil industry is critical to profitability. An error of only 0.1% is worth many millions of US dollars in the bottom line.

Only when the data quality is understood, can proper and cost-effective business decisions be made.

Data Quality and Business Decision Examples

Data quality assessment is required for business, research and development decisions.

No test data is useful without an accompanying data quality assessment. Uncertainty analysis is the foundation for quantitative data quality assessment. Objective uncertainty estimates are required for custody transfer, measurement and test data to be useful.

It is a fundamental requirement for competent engineering or scientific activities to have a comprehensive, robust data quality metric, its measurement uncertainty, assigned to data developed and reported.

Significant risk reduction results when a data quality metric, uncertainty, accompanies data secured or provided.

Great financial risk will result when decisions are based on data that does not have an accompanying uncertainty analysis, that is, a quantitative data quality assessment. No test data should be either reported or used without knowledge of its quality, its measurement uncertainty. Violating this precept causes undue risk of incorrect business decisions.

Example 1: Product Durability Improvement

Consider a company that works to improve the durability of its product. The current product lasts 12 months but there is demand from customers for a product that will last 36 months. The engineers and scientists of the company have developed improvements that they believe will increase product durability to 37 months. Testing data supports this conclusion. The company then goes into production with its improved product.

However, no uncertainty analysis was done on the 37-month lifetime of the new product. In fact, errors of the testing process when propagated into product life suggest the 95% confidence uncertainty of the 37-month lifetime is ±2.5 months. This means that the actual new lifetime of the product could be anywhere from 34.5 months to 39.5 months, 19 times out of 20. This is too large a band on which to send a product into production. The result will be that this company has great risk that it will have many unhappy customers, possibly be the target of liability court cases, and, in any case, incur major increased, unbudgeted costs.

Example 2: Custody Transfer Costs

In the gas and petroleum industries, many millions of US dollars paid and received depend on the correct knowledge of the custody transfer processes. Consider an oil producer that sells a refinery 500,000 bbl of crude oil but the refinery, based on its measurements of the pipeline delivery, claims that only 495,000 bbl were received. How can this huge cost error be corrected? What are the risks to the producers and the refiners? A clear uncertainty budget process will identify the sources of major errors and provide guidance on how to correct them. If this kind of data quality assessment is not done, producers, refiners and distributors will be at great financial risk.

Example 3: Need for Purchase of Improved Equipment

An instrumentation or control system sales engineer visits a company's facilities and recommends that the company purchase his "improved" equipment, instruments and/or control systems. The purchase requires that installation of the new equipment will improve the profitability of the company. However, without a comprehensive uncertainty analysis of the processes in the company, there can be no assurance that the new instruments or control systems will have a significant effect. The new devices must impact the areas of the company's processes that have the largest instances of poor measurement or control. Uncertainty analysis assists in assessing the areas needing the most improvement.

Both the instrument or control sales engineer and the company owning the processes must understand the impact of uncertainty to better assess where improvements are warranted. The company manager and/or engineer must understand this to make economically valid purchases. The sales engineer must understand this to add credibility to his recommended sales.

Example 4: Need for External Calibration

The decision as to whether or not have an instrument sent to a calibration laboratory rests on the assumption that the laboratory calibration is needed and sufficiently accurate (low enough uncertainty). Here, uncertainty analysis assists both the instrument owner and laboratory operator. The instrument owner in possession of an up-to-date uncertainty analysis of the use of the instrument will know what level of certification to pay for. The laboratory owner with a valid uncertainty analysis of his facilities will have the objective evidence supporting his claim to fill the need of the instrument owner. Both individuals need robust, objective uncertainty estimates to accomplish their job, company profitability.

The Bottom Line

Assuring that engineers, managers and executives alike understand the impact of measurement uncertainty is the only way to understand the risks a company faces in its profitability plans.

- Objective methods exist for data quality assessment that are accepted throughout the world.

- Measurement uncertainty analysis is the best way to objectively assess data quality.

- Data quality has a significant impact on a company's profitability.

- Executive and management level understanding of the impact of measurement uncertainty is critical to the financial success of a company.

4-10. The Timing of an Uncertainty Analysis

There are two major times an uncertainty analysis should be done. One is pretest, or before a test is even run. The other is post test, or after the test results are in for analysis. Pretest uncertainty analysis tells the experimenter whether or not the planned test is adequate to observe the effect of interest. Post test uncertainty analysis provides a diagnostic test that reveals the value of the data for decision making.

Pretest Uncertainty Estimates

To the uninitiated experimenter, the first response to the suggestion that a pretest uncertainty analysis is needed is, "Why bother?" What is often missed is that every experimenter does an uncertainty analysis. It is usually after the test and suggests that either the uncertainty is low (results match predictions) or that it is large (results refute predictions). Neither of these approaches to uncertainty analysis is appropriate.

A pretest uncertainty analysis is a test planning tool that, when done correctly, yields two major kinds of information:

A. Is the experiment worth running? Will the expected test result be observable considering the testing errors?

B. If the uncertainty is too large, what should be done to reduce the error so that the test result might be observable with confidence?

It is foolish to design a test to demonstrate a 0.5% improvement in compressor efficiency, for example, and then use a measurement system that has an uncertainty of ±1.0%. Any test result observed has a high likelihood of having been caused by the errors of the measurement system and not the design change intended to improve the compressor efficiency. Therefore, the primary question to be answered by a pretest uncertainty analysis is: *Is the predicted uncertainty good enough?* Or, is the test result clearly observable in the fog of the measurement errors?

Numerous responses are received from an experimenter, who is low on budget and who realizes that an uncertainty analysis may be appropriate but may not want to admit it. Such experimenters fall into five categories.

A. Ostrich—"I don't want you guys putting fuzz bands on my data." The fallacy here is that the "fuzz bands" are already there and all that an uncertainty analysis will do is make them visible and addressable. Ignoring them, like the ostrich, will only yield major problems down the road.

B. Doubter—"I don't believe in all that uncertainty mumbo jumbo." This type of individual needs an education. Uncertainty analysis is a simple approach to putting objective limits on the validity of data. It contributes to an understanding of the test result.

C. Indifferent—"This is not my favorite subject." This experimenter is more worried about conducting the test than about what value may be realized from the data. It is important to understand that the data have value only if the effects being searched for exceed the uncertainty. It should be every experimenter's "favorite subject."

D. Converted—"After talking to you guys, I get the impression that if I don't start worrying about validity I'm going straight to hell." This individual has seen that validity, or uncertainty analysis, is critical to a successful test program.

E. Biblical—"A false balance is an abomination to the Lord; but a just weight is His delight" (Ref. 4). Uncertainty analysis is intended to provide information on just how "just" the weight is. Its proper analysis and application are central to a decision process based on experimental results.

Pretest uncertainty analysis tells the experimenter, going in, what to expect. If the uncertainty is too large, it points out those areas that most merit improvement (where the money should be spent).

Example 4-3:

A pretest uncertainty analysis suggests that the expanded uncertainty in the delta pressure measurement transducer across an orifice is ±0.2 psid. The purpose of the experiment is to measure an increase in pressure of 0.15 psid. What is recommended?

What should be done is to not run the experiment! An observed change of 0.15 psid might be solely due to the error. The various error sources should be reviewed to determine the major error contributors (or drivers) and improvements accomplished there before any test of this type is run.

If the systematic standard uncertainty is too large, the following actions could be taken:

A. Improve the calibration with a better procedure or standards.

B. Utilize independent calibrations. This will help average out systematic error in one method.

C. Utilize concomitant variables. If the flow can also be measured with a turbine meter, even an expanded uncertainty of ±0.2 psid on the transducer might be adequate for the experimenter to observe a 0.15 psid change. (More on this later in Unit 5.)

D. In-place calibration. The errors that accumulate in a calibration hierarchy can be bypassed sometimes if the calibration standards are brought out to the test site.

If the random standard uncertainty is too large, the following actions could be taken.

A. Take a larger data set. This will often average out the random errors, but it is important to take the large data set over all the variables that contribute most to the random uncertainty. It will do no good, for example, to push the button of the data system, thus acquiring reams of data to average, if the biggest random errors are day-to-day variations. In that case, many days of data are needed.

B. Better instrumentation. Instrumentation that is more precise will often help, but this is frequently an expensive solution.

C. Redundant instrumentation. As discussed earlier, two identical instruments could measure the same parameter or variable, and the resultant random uncertainty would go down from that of a single instrument by a factor of the square root of two.

D. Other techniques too detailed to discuss here, such as moving averages, data smoothing, regression analysis, etc. Consult a statistician for help with these.

E. Improve the test design. This is often the easiest solution. The acquisition of the data, more data, and altering the concentration of data can all help. In the latter case, for example, if the real interest in a test result is the 90% level of a transducer, most of the calibration points should be in that region, not over the whole range of the transducer.

Pretest uncertainty analysis should give the experimenter an idea of where to head. Post test uncertainty analysis tells the experimenter where he or she has been.

Post Test Uncertainty Estimates

There are several reasons for post test uncertainty analysis:

A. For future pretest uncertainty analysis. Does the observed data scatter match the predicted random standard uncertainty? If not, something needs to be learned about the experimental method before it is tried again.

B. Competitive method analysis. Do expanded uncertainties overlap? They should. If they don't, one or both of the uncertainty analyses is wrong and should be investigated.

C. Specifications satisfied? Does the test result with its expanded uncertainty provide data on which decisions can be based?

As an aside, the question should be asked, "Is the experimental systematic error observable in the test result?" This will be a problem for the student.

Post test uncertainty analysis tells the experimenter where he or she has been and what to do better the next time. Without it, the experimenter is sailing on a sea of uncertainty with neither a compass (knowledge of systematic uncertainties) nor a rudder (knowledge of random uncertainties).

4-11. Conclusion

It should be concluded that uncertainty is a part of all test data, and no test data should be considered without also considering its uncertainty.

References

1. ANSI/ASME PTC 19.1-2006, *Performance Test Code, Instruments and Apparatus, Part 1, Measurement Uncertainty.*

2. Ibid., p. 11.

3. Ibid., p. 31.

4. *The Bible*, Proverbs 11.1.

5. *International Standards Organization (ISO) Guide to the Expression of Uncertainty in Measurement*, 1993.

6. NIST Technical Note 1297, 1994 Edition, *Guidelines for Evaluating and Expressing the Uncertainty of NIST Measurement Results.*

Exercises

4-1.   Terminology Problem

In your own words, define the difference between these three pairs of symbols:

a. ε and $2s_{\overline{X}}/\sqrt{N}$

b. β and B

c. δ and U

4-2.   Pooling Review Problem

Consider the following three standard deviations:

$$s_{X,1} = 40.0, \nu = 12$$
$$s_{X,2} = 50.0, \nu = 15$$
$$s_{X,3} = 30.0, \nu = 7$$

a. What do $s_{X,1}$, $s_{X,2}$, and $s_{X,3}$ represent?

b. What is the best estimate of the population true standard deviation, σ, in the above standard deviations?

c. Calculate the best estimate of the population true standard deviation.

d. What are the degrees of freedom associated with (c)?

4-3.   Delta Problem

Consider the following delta readings between two identical measurement systems:

System 1	System 2
10.3	11.2
9.5	10.5
12.8	13.4

a. What is the standard deviation for System 1 and for System 2?

b. Do the results in (a) represent the correct expected standard deviation for the systems?

c. Calculate the deltas between the two systems, assuming they are measuring the same process.

d. Calculate the standard deviation of the deltas.

e. Obtain the standard deviation for either system (assuming they are identical) from (d).

f. What is learned by comparing (a) and (e)?

4-4.   Interfacility Results Problem

Consider that seven laboratories participate in a round robin experiment. The average result for an orifice calibration is 12.0 lb/s for given flow conditions. The standard deviation between laboratories is ±0.2 lb/s.

a.   What does the value of 12.0 lb/s represent?

b.   What is the laboratory-to-laboratory random standard uncertainty?

c.   What does (b) represent from the view of the average for all the laboratories?

d.   What does (b) represent from the point of view of each laboratory?

4-5.   Systematic Standard Uncertainty Impact Problem

Consider the elemental systematic standard uncertainties of 1, 2, 3, 10, 20, and 30 psi.

a.   Using Equation 4-9, obtain the systematic standard uncertainty, b_R.

b.   Is (a) a reasonable estimate of the systematic standard uncertainty? Why or why not?

c.   Using Equation 4-10, obtain the systematic standard uncertainty for a result, b_R.

d.   Is (c) a reasonable estimate of the systematic standard uncertainty? Why or why not?

e.   Using Equation 3-7, obtain the systematic standard uncertainty.

f.   Is (e) a reasonable estimate of the systematic standard uncertainty? Why or why not?

g.   Of (a), (c), and (e), which is the best estimate of the systematic standard uncertainty for a result?

4-6.   Back-to-Back Calibration Problem

The expanded uncertainty in a heat transfer measurement is:

Systematic standard uncertainty of the measurement (b_R) = 1%

Random standard uncertainty of the measurement $(s_{\bar{X}, R})$ = 1%

$U_{95} = \pm 2[(b_R)^2 + (s_{\bar{X}, R})^2]^{1/2} = \pm 2.8\%$ (large sample sizes)

Half of the systematic and random standard uncertainties are due to calibration; that is,

$$b_{CAL} = (1/2)^{1/2}\%, \qquad\qquad b_{\bar{X}-CAL} = 1/2)^{1/2}\%$$

a.   For a single test, single calibration, what are the calibration standard uncertainties, systematic and random?

b.   For many tests $(d.f. > 30)$ over a period of years involving many calibrations, what are the calibration standard uncertainties?

c.   For a back-to-back comparative test where the objective is to determine the difference between two tests with the same instrumentation calibrated once, what are the calibration standard uncertainties?

d.   After a period of years, an inspection of the calibration histories shows that this instrumentation drifts with time in a cyclical manner. What would you do about this? Would the above estimates change?

4-7.   Observable Systematic Error Problem

After a series of tests in an oil refinery, it was determined that the custody transfer of crude from the tanker to the refinery tanks was complete and that the total flow was 1.5 million barrels ±0.5%. This represents two standard deviations for the several flow-time integrations that add up to the total.

a.   What is the expanded uncertainty of the flow measured? (Trick question.)

b.   Is the effect of the systematic error observable in the test results? Why or why not?

c.   What is the risk involved in calling the ±0.5% the flow uncertainty?

4-8.   Compressor Uncertainty Problem

As a data validity engineer, you conducted a pretest uncertainty analysis for a compressor rig test. The results (for many degrees of freedom) were:

Systematic standard uncertainty $(b_R) = 0.4\%$

Random standard uncertainty $(s_{\bar{X}, R}) = 0.6\%$

$$U_{95} = 2[(b_R)^2 + (s_{\bar{X}, R})^2]^{1/2} = \pm 1.4\%$$

The compressor project engineer responsible for the test predicts that the efficiency measured for the new design will be better than the old by 0.9%.

a.   What would you advise?

b.   After the test, plots of data for 10 repeated tests show a scatter band of about ±1.0%. Is this consistent with your analysis?

c.   What can be done to reduce the random standard uncertainty?

d.   What can be done to reduce the systematic standard uncertainty?

4-9.   Historic Judgment Problem:

You are doing a pretest measurement uncertainty analysis of a burner rig, a single test with a single calibration. The test objective is to determine burner efficiency. How would you use the following information?

a.   There is historic data that indicates there is some stand-to-stand variation, $s_X = \pm 0.1\%$.

b.   Same as (a) but there is no data. Still, from engineering judgment, you suspect stand-to-stand differences.

c.   The third shift crew measures efficiencies 0.2% higher than the first and second shifts crews based on data over the past year.

d.   Some old data indicates the time-to-time effect, the same rig tested over time, should show a variation of $s_X = \pm 0.3\%$.

e.   Production data indicates there is probably real variation, not error, on the order of $s_X = \pm 0.4\%$.

4-10.   "Why measure that way" Problem

An automobile emissions test is run to determine whether or not the car exceeds the Environmental Protection Agency certification limits of 100 parts per million (ppm) carbon monoxide, 100 ppm hydrocarbons and 100 ppm nitrogen oxides. Results of the two power levels tested (both of which must be under the limits to pass the car) are as follows along with their uncertainties:

Gas	Idle Power	Max Power	95% Conf. Uncertainty
CO	105 ppm	22 ppm	±6 ppm
HC	98 ppm	3 ppm	±7 ppm
NO_X	8 ppm	150 ppm	±10 ppm

a.   Assume you are the Regulating Agency; decide whether or not a power level gas combination passes or fails certification and enter P or F as appropriate below and explain your reasoning.

	Idle	Max	Reasoning
CO			
HC			
NO_X			

b.   Now assume you are the Automobile Manufacturer and fill out the chart again with P or F for pass or fail respectively and explain your reasoning.

	Idle	Max	Reasoning
CO			
HC			
NO_X			

c.   Are the charts different? Where? Why?

d.   If you are the manufacturer, what can you do with your equipment to possibly pass these tests?

e.   What is interesting about the max-HC data and the idle-NO_X data?

Unit 5:
Uncertainty (Error)
Propagation

UNIT 5

Uncertainty (Error) Propagation

In this unit, you will learn the need and the methods for propagating measurement uncertainties into their effect on the test result. A measurement uncertainty analysis is not complete after the measurement uncertainties of the individual measurements are complete. Their effect on the test result must be evaluated. This unit provides the methods for that evaluation.

Learning Objectives—When you have completed this unit you should:

A. Understand the need for uncertainty (or error) propagation.

B. Understand the theoretical background of uncertainty propagation.

C. Be able to propagate elemental systematic and random standard uncertainties into a test result.

5-1. General Considerations

It is not enough to evaluate the magnitudes of the elemental standard uncertainties or the magnitudes of the standard uncertainties of the individual measurements such as temperature and pressure. Their combined effect on the test result must be determined. In previous units in this book, the U_{95} model has been presented as the method for the calculation of measurement uncertainty. In every case, the uncertainties being combined were all of the same units—all temperature, all pressure, or all flow, for example. If a measurement of temperature and pressure yields, through calculation, a flow, combining the temperature and pressure elemental uncertainties cannot be done simply with the methods already presented.

In this unit, you will learn to propagate the effects of the systematic and random standard uncertainties for each measurement (i.e., temperature or pressure) into a test result calculated from several measurements.

5-2. The Need for Uncertainty Propagation

It should be noted that both errors and the estimates of their limits, uncertainties, can be propagated. However, this unit deals only with uncertainty propagation since it is that which results in an understanding of the value of a calculated result. That is, there is a need to calculate the uncertainty of a result that arises from uncertainties in the terms combined to compute that result.

Combining elemental standard uncertainties from sources with different units is best accomplished by first converting them to the same units. Usually the simplest and best way to do that is to propagate them into result units. When the standard uncertainties are all expressed in result units, their combination is as simple as shown in previous units of this book. All uncertainty propagation formulas and methods have the following basic assumption of application: Identical units only are used.

For example, suppose there are two random standard uncertainties whose magnitudes and units are 2.0°F and 3.0°F. Their combined effect on a test result is computed using Equation 3-3:

$$s_{\overline{X}, R} = [\sum (s_{\overline{X}, i})^2]^{1/2} \qquad\qquad (5\text{-}1)$$

Remember that:

$$s_{\overline{X}, i} = [s_{\overline{X}, i}(N_i)]^{1/2}$$

Therefore, for this example:

$$s_{\overline{X}, R} = \{[(2.0^2)(°F)^2] + [(3.0^2)(°F)^2]\}^{1/2}$$
$$= \{(13)(°F)^2\}^{1/2}$$
$$= 3.6°F$$

Note that the random standard uncertainties and the units calculate properly. This is adequate treatment where the combination is linear. Nonlinear equations used to compute results must utilize uncertainty propagation, as shown below for a linear example where two different units are used for the two random standard uncertainties.

Consider the case where a result is obtained from the combination of two uncertainty sources: temperature and pressure. Assuming the temperature random standard uncertainty is 2.0°F and the pressure random standard uncertainty is 3.7 psia, an attempt is made to use Equation 5-1 as follows:

$$s_{\overline{X}, R} = [\textstyle\sum (s_{\overline{X}, i})^2]^{1/2}$$

In this case, the following is obtained:

$$s_{\overline{X}, R} = \{[(2.0^2)(°F)^2] + [(3.7^2)(psia)^2]\}^{1/2}$$

As can be seen, the 2.0 and 3.7 might be able to be combined, but what can be done with the °F and psia portions of the above equation? It is necessary to propagate the effects of the two random standard uncertainties into the test result.

Postulating that the result equation has the form of Equation 5-2, the uncertainty propagation may proceed:

$$R = XY \tag{5-2}$$

where X is temperature in degrees F and Y is pressure in psia.

The propagation of uncertainty in this result is detailed later under "Taylor's Series for Two Independent Uncertainty Sources."

5-3. Theoretical Considerations: Methods for Uncertainty Propagation

There are three well-known methods for uncertainty propagation: Taylor's Series (Refs. 1, 2), "dithering," and Monte Carlo simulation. The first method is by far the most commonly utilized; however, it is the aim of all three to convert the units of the uncertainty sources (systematic and random) to units of the test result. This is done in each case by determining the *sensitivity* of the test result to uncertainty in each source. This sensitivity is sometimes called the *influence coefficient*.

The most common method, Taylor's Series approximation, will be covered in detail. Its derivation, however, and the statistical detail will not be covered here; Refs. 1 and 2 both do that adequately. In addition, numerous statistical texts deal with Taylor's Series approximations.

It should be noted that in this writer's experience, it has always been acceptable to determine the uncertainty of a test result by using the "college text book," closed form expression for the calculation of the result in question. As an example of how this works, consider the computation of compressor efficiency by a company that develops and markets compressors. The calculation is complicated by many correction factors.

However, the uncertainty itself may be estimated by considering only four factors. The four are the inlet and exit, temperature and pressure. With the college textbook closed form of the equation for efficiency, a fully adequate calculation of the measurement uncertainty can be done. Although this simpler equation for efficiency is not adequate for a company marketing their compressor performance, it is fully adequate for uncertainty estimation. Remember, uncertainty analysis is an estimation process. Closed form equations for the results are almost always adequate.

Taylor's Series Uncertainty Propagation

Physical Explanation for the Taylor's Series Uncertainty Propagation

Determination of a quantity that is expressed by a functional relationship of one or more measured variables often requires a knowledge of the effect of the uncertainty in the measurement of the variables on the computed function. One approach to evaluating the effects of these uncertainties is called Taylor's Series uncertainty propagation (Ref. 3), in which only first-order terms are considered significant.

Briefly, this approach to uncertainty propagation proceeds as follows:

Given the function

$$X = F(Y) \tag{5-3}$$

it is desired to know the expected uncertainty in X due to a known uncertainty in Y. One approach is to realize that if the slope of X with reference to Y at some point Y_i could be determined, it would provide an estimate of the effect that small changes in Y about Y_i, (ΔY), would have on X about X_i, (ΔX). This slope is easily evaluated by differentiating X with respect to Y and evaluating the resulting expression at Y_i, that is,

$$\text{Slope in } X \text{ at } Y_i = \frac{\partial F}{\partial Y} \tag{5-4}$$

The sign of $(\partial F/\partial Y)$ is an indication of the effect of the sign of the Y uncertainty; that is, if $(\partial F/\partial Y)$ is negative, a plus Y uncertainty results in a decrease in the value of X.

The effect of this slope is dependent on the uncertainty in Y, ΔY at Y_i; in fact, it is seen that:

$$(\Delta X)^2 = (\partial F/\partial Y)^2 (\Delta Y)^2 \tag{5-5}$$

Here $(\Delta X)^2$ and $(\Delta Y)^2$ are easy to recognize physically as the variance in X and Y, respectively. Equation 5-5 is the basis of the Taylor's Series uncertainty propagation.

When X is a function of more than one independent variable, that is,

$$X = F(Y,Z) \tag{5-6}$$

the statistical procedure is to sum in quadrature:

$$(\Delta X)^2 = (\partial F/\partial Y)^2(\Delta Y)^2 + (\partial F/\partial Z)^2(\Delta Z)^2 \tag{5-7}$$

(A similar extension of the equation is used for functions of more than two variables.)

Physically, this could be thought of as the evaluation of the change in two orthogonal variables, Y and Z, with independent uncertainty sources, ΔY and ΔZ, and their effect, ΔX, on the dependent variable X through the use of the Pythagorean theorem.

Note that the signs of $(\partial F/\partial Y)$ and $(\partial F/\partial Z)$ are known but then are lost due to squaring and due to incomplete knowledge about the direction of the Y and Z errors. The relative magnitudes of $(\partial F/\partial Y)$ and $(\partial F/\partial Z)$ are an indication of how careful the experimenter must be in the measurement of Y and Z in the regions of Y_i and Z_i.

The proper evaluation of Equation 5-7 requires that everything be in the same units and that ΔX be independent of ΔY. The simplest approach is to evaluate the expression in absolute units and then convert the resulting ΔX into a percent uncertainty after summing the individual uncertainties.

Taylor's Series for Two Independent Uncertainty Sources

Let's begin with an expression for a test result of

$$R = F(X,Y) \tag{5-8}$$

where:

R = the test result that has a continuous partial derivative and is itself continuous about the point (X,Y)

X = the first variable that has a continuous error distribution (usually normal) in the region of the point (X,Y)

Y = the second variable that has a continuous error distribution (usually normal) in the region of the point (X,Y)

The constraints on R, X, and Y are shown in the description of the variables of Equation 5-8.

The Taylor's Series equation for propagating the effects of uncertainty in X and Y on the result R for independent uncertainty sources is:

$$(U_R)^2 = (\partial R / \partial X)^2 (U_X)^2 + (\partial R / \partial Y)^2 (U_Y)^2 \tag{5-9}$$

where:

$U_R =$ an uncertainty in the result, either systematic or random

$U_X =$ an uncertainty in the measured parameter X, either systematic or random

$U_Y =$ an uncertainty in the measured parameter Y, either systematic or random

$\partial R / \partial X =$ the influence coefficient that expresses the influence an uncertainty in X will have on result R at level X_i

$\quad = \theta_X$, an alternate notation

$\partial R / \partial Y =$ the influence coefficient that expresses the influence an uncertainty in Y will have on result R at level Y_i

$\quad = \theta_Y$, an alternate notation

Note that although each "U" may be either systematic or random standard uncertainty, in each use of Equation 5-9 all the "U" must be one or the other. That is, all "U" in each equation must be either systematic or random.

Using Equation 5-9, the effects of an uncertainty in either X or Y may be impressed upon result R. The influence coefficients, evaluated at the X and Y levels of interest, will convert any standard uncertainty, systematic or random, into an uncertainty in the result, R. The influence coefficients are actually the slopes of the uncertainty surface at (X_i, Y_i) with respect to either measured parameter, X or Y. The variable in Equation 5-9 may be replaced by either systematic standard uncertainty, b, or the random standard uncertainty, $s_{\overline{X}}$. (Remember, $s_{\overline{X}} = s_X / \sqrt{N}$.) Thus, the effect of the combination of uncertainties from many sources can be computed with an expression such as Equation 5-9 that is specific for two uncertainty sources. The general expression for N uncertainty sources is:

$$(U_R)^2 = \sum_{i=1}^{N} [(\partial R / \partial V_i)^2 (U_i)^2] \tag{5-10}$$

where:

$V_i =$ the ith measured parameter or variable

U_i =   the uncertainty (systematic or random) associated with the ith parameter

In addition, the use of the sensitivity coefficients permits the root sum squaring of the uncertainty components since, once multiplied by their individual influence coefficients, each is in the units of results. In other words, the square root of Equation 5-10 is the uncertainty in R due to uncertainties in the measured parameters that go to make up R. (In case you are interested, the units of each influence coefficient are [(results)/ (variable of interest)].) This capability of the influence coefficients is central to uncertainty propagation, which could not be accomplished without it.

Remember that one assumption of this approach is that the uncertainty sources in X and Y are independent.

Step-by-Step Approach to Simple Uncertainty Propagation, Two Measured Parameters, Independent Uncertainty Sources

To get an idea of how this approach works, consider the result expression given by Equation 5-2:

$$R = XY$$

In this expression, the result, R, is dependent on the measurement of X (in, for example, degrees) and Y (in, for example, psia).

The uncertainty propagation steps to follow are:

A.   Write the closed-form expression for the result as has been done in Equation 5-2.

B.   Derive the influence coefficients (or sensitivities) with partial differentiation.

C.   Evaluate the magnitudes of the influence coefficients at the point of interest, (X_i, Y_i).

D.   Fill in Equation 5-9 with the appropriate values and calculate the uncertainty in R caused by the uncertainties in X and Y.

E.   Compute the percent uncertainty in R last, at the R level of interest.

Now for the example of Equation 5-2, the sensitivities are as follows:

$$\theta_X = \partial R / \partial X = Y \tag{5-11}$$

$$\theta_Y = \partial R / \partial Y = X \tag{5-12}$$

Assuming the systematic standard uncertainty for X is 2.0°F and the systematic standard uncertainty for Y is 3.7 psia, the following is obtained using Equation 5-10 in the region of (150°F, 45 psia):

$$\begin{aligned}
b_R &= [(Y^2)(b_X)^2 + (X^2)(b_Y)^2]^{1/2} \\
&= [(45^2)(2)^2 + (150^2)(3.7^2)]^{1/2} \\
&= [8100 + 308,025]^{1/2} \\
&= 562°\text{F-psia}
\end{aligned} \tag{5-13}$$

(Note in the above that we are using all the systematic standard uncertainties that easily fit our U_{95} uncertainty model.)

Since the R level of interest here is (150°F)(45 psia) = 6750°F-psia, the percent uncertainty in R is (562/6750)(100) = 8.3%.

Note that the percent systematic standard uncertainties in X and Y were 1.3% and 8.2%. Were these to be simply root-sum-squared, the result uncertainty in R would have been 8.3%, luckily the right answer. Proper uncertainty propagation in absolute units is an important step in uncertainty analysis. It is often not possible to merely combine percent standard uncertainties as it was for this simplified example.

Note that it is possible to express all these uncertainty propagation formulas such that the units of the uncertainties are relative, that is, percent. They will work then, too; however, it can get confusing. I recommend that all uncertainty propagation be done in absolute or engineering units and no conversion to relative units be done until after the final result uncertainty is determined.

Note too that Equation 5-13 may be used to identify the uncertainty source of significance. It reveals the major contributor to uncertainty in the result as Y, not X. It would not pay the experimenter to work hard and spend money improving the temperature measurement. Almost all the uncertainty is caused by the pressure measurement uncertainty. This is where the improvements should be concentrated if the predicted uncertainty in the result of 8.3% is unacceptable. (This is called an uncertainty budget.)

(Note also that here is the usual place where Pareto charts are used to make clear the major error sources. Most statistics texts discuss Pareto charts.)

Taylor's Series for Two Nonindependent Uncertainty Sources

It is sometimes the case that the error sources are not independent. That is, one is correlated to some extent with another. In this case, it is not possible to use the uncertainty propagation equations just shown. Consider the example of Equation 5-2:

$$R = XY$$

where the errors and hence the uncertainties in X and Y are not independent.

The expression for uncertainty propagation, Equation 5-10, does not apply then. That expression was for independent uncertainty sources in X and Y. However, we are saved by the mathematician again; the more general formula for uncertainty propagation with a Taylor's Series for two variables is:

$$(UR)^2 = \sum_{i=1}^{N} \{[(\partial R/\partial V_i)^2(U_i)^2] + [\text{all cross-product terms}]\} \qquad (5\text{-}14)$$

where V_i is the ith measured parameter or variable. Cross-product terms have two variables in each and are generically shown as:

$$[2(\partial R/\partial V_a)(\partial R/\partial V_b)r_{ab}U_aU_b]$$

where:

a and b	=	the complete set of all the combinations of two variables possible
U_a and U_b	=	the uncertainty in variables a and b
r_{ab}	=	the sample correlation of error in a on error in b

Note that

$$r_{ab} = \frac{\{N[\sum(a_ib_i)] - (\sum a_i)(\sum b_i)\}}{\left\{[N(\sum a_i^2) - (\sum a_i)^2][N(\sum b_i^2) - (\sum b_i)^2]\right\}^{1/2}} = \frac{s_{ab}}{s_a s_b} \qquad (5\text{-}15)$$

and that

$$r_{ab}U_aU_b = U_{ab}$$

where U_{ab} is the covariance of a on b. (Stick with us; this gets simplified soon.)

Here, too, the variables a and b represent all combinations of two variables from the general expression for the uncertainty propagation formula, Equation 5-14. A complete set of these terms for two variables is shown in Equation 5-16 and for three variables in Equation 5-19.

Step-by-Step Approach to Simple Uncertainty Propagation, Two Measured Parameters, Nonindependent Error Sources

To get an idea of how this approach works, consider Equation 5-2:

$$R = XY$$

In this expression, the result, R, is dependent on the measurement of X (in, for example, degrees) and Y (in, for example, psia).

The error propagation steps to follow are:

A. Write the closed-form expression for the result as has been done in Equation 5-2.

B. Derive the influence coefficients (or sensitivities) with partial differentiation.

C. Evaluate the magnitudes of the influence coefficients at the point of interest, (X_i, Y_i).

D. Fill in Equation 5-14 with the appropriate values and calculate the uncertainty in R caused by the uncertainties in X and Y.

E. Compute the percent uncertainty in R last, at the R level of interest.

Now for the example of Equation 5-2, the sensitivities remain as before and are, as in Equations 5-11 and 5-12:

$$\theta X = \partial R/\partial X = Y$$

$$\theta Y = \partial R/\partial Y = X$$

Applying Equation 5-14 and assuming the correlation between the errors in X and Y is 1.0, the following is obtained:

$$(U_R)^2 = (\partial R/\partial X)^2(U_X)^2 + (\partial R/\partial Y)^2(U_Y)^2 + 2(\partial R/\partial X)(\partial R/\partial Y)r_{XY}U_XU_Y$$
$$= (\partial R/\partial X)^2(U_X)^2 + (\partial R/\partial Y)^2(U_Y)^2 + 2(\partial R/\partial X)(\partial R/\partial Y)U_{XY} \qquad (5\text{-}16)$$

where:

$U_R =$ a standard uncertainty in the result, either systematic or random

$U_X =$ a standard uncertainty in the measured parameter X, either systematic or random

$U_Y =$ a standard uncertainty in the measured parameter Y, either systematic or random

$r_{XY} =$ the correlation of the error in X on the error in Y

$U_{XY} =$ the covariance of X on Y

As shown in Ref. 4, U_{XY}, the covariance, equals the product of the perfectly correlated uncertainties; usually this means $b'_X b'_Y$, where b'_X and b'_Y are the portions of the X and Y systematic standard uncertainties for which their errors are perfectly correlated.

It is instructive to see how this will work for clearly correlated systematic standard uncertainty terms. Consider the Equation 5-27 at this time.

$$b_{\eta,c} = \{[(\Theta_{P1})(b_{P1})]^2 + [(\Theta_{P2})(b_{P2})]^2 + [(\Theta_{T1})(b_{T1})]^2 + [(\Theta_{T2})(b_{T2})]^2$$
$$+ 2\Theta_{T1}\Theta_{T2}[(b'_{T1})(b'_{T2})/(b_{T1})(b_{T2})]*(b_{T1})(b_{T2})\}^{1/2} \qquad (5\text{-}27)$$

We will deal with this equation in more detail later in this chapter but for now, we want to see how this equation will reduce to the form in which it appears on page 131.

The term in 5-27 above, $[(b'_{T1})(b'_{T2})/(b_{T1})(b_{T2})]*(b_{T1})(b_{T2})$, is actually the covariance term and could have been written as b_{T1T2}. In addition, the term $[(b'_{T1})(b'_{T2})/(b_{T1})(b_{T2})]$ is actually the sample correlation coefficient, r. (Note this is all covered in some detail in Reference 4.)

As can be see in the product of these two terms, the $(b_{T1})(b_{T2})$ in the denominator cancels with the $(b_{T1})(b_{T2})$ in the numerator leaving us with only $(b'_{T1})(b'_{T2})$, or the product of only the correlated pairs of systematic uncertainties, here noted with the prime (').

When we deal with correlates systematic errors, we now only have to carry along all the pairs of the correlated systematic standard uncertainties. We do not have to calculate correlation coefficients or covariances. We therefore have the simplified form of Equation 5-27 as shown on page 131.

From Equation 5-16 and remembering that the correlated systematic standard uncertainties are ±2.0°F for X and ±3.7 psia for Y and the levels of interest for X and Y are 45°F and 150 psia, respectively, the following is obtained.

$$
\begin{aligned}
b_R &= [(Y^2)(b_X)^2 + (X^2)(b_Y)^2 + 2YXb_{XY}]^{1/2} \\
&= [(Y^2)(b_X)^2 + (X^2)(b_Y)^2 + 2YXb'_Xb'_Y]^{1/2} \\
&= [(45^2)(2)^2 + (150^2)(3.7^2) + 2(150)(45)(2.0)(3.7)]^{1/2} \\
&= [8100 + 308{,}025 + 99{,}900]^{1/2} \\
&= 645°F\text{-psia}
\end{aligned}
\tag{5-17}
$$

Since the R level of interest here is $(150°F)(45 \text{ psia}) = 6750°F$-psia, the percent standard uncertainty in R is $(645/6750)(100) = 9.6\%$. This is a significantly larger uncertainty than would have been obtained if the error dependency had been ignored.

Additional Example of Step-by-Step Instructions

Consider now the following result expression:

$$R = X/Y$$

From Equations 5-11 and 5-12, the sensitivities are:

$$\theta_X = \partial R/\partial X = 1/Y = 0.022222$$

$$\theta_Y = \partial R/\partial Y = -X/Y^2 = -0.074074$$

Applying Equation 5-16 and again remembering that the correlated systematic standard uncertainties are ±2.0°F for X and ±3.7 psia for Y and the levels of interest for X and Y are 45°F and 150 psia, respectively, the following is obtained:

$$
\begin{aligned}
b_R &= [(1/Y)^2(b_X)^2 + (-X/Y^2)^2(b_Y)^2 \\
&\quad = 2(1/Y)(-X/Y^2)b'_Xb'_Y]^{1/2} \\
&= [(0.022222^2)(2^2) + (-0.074074^2)(3.7^2) \\
&\quad = 2(0.022222)(-0.074074)(2.0)(3.7)]^{1/2} \\
&= [0.0019753 + 0.0751164 - 0.0243619]^{1/2} \\
&= 0.23°F/\text{psia}
\end{aligned}
$$

Since the R level of interest here is $(150 \text{ psia})/(45°F) = 3.33°F/\text{psia}$, the percent standard uncertainty in R is $(0.23)(100)/(3.33) = 6.97\%$. This is obviously a significantly smaller standard uncertainty than would have been obtained if the error dependency had been ignored. Ignoring the dependency or correlation would have resulted in using only the first two terms of the equation, that is, $(0.0019753 + 0.0751164)^{1/2}(100)/(3.33) = 8.34\%$. Indeed, some of the uncertainty in the result disappears when there is perfect correlation between the two error sources for this example.

It should be noted that the above methods for nonindependent uncertainty sources may be applied to uncertainty sources that are partially dependent ($|r| > 0.0$) although not completely nonindependent ($|r| = 1.0$). Note too that r, the correlation coefficient, may vary from -1.0 to $+1.0$. A negative correlation coefficient implies that the second variable is inversely correlated with the first. This means that an increase in the second variable, or uncertainty source, will produce a decrease in the first. A positive correlation coefficient means that the second variable is directly correlated to the first. In this case, an increase in the second variable will result in an increase in the first variable as well. Partially correlated errors require the use of r in the propagation equations.

Application to Three Variables

It is instructive to see the equations that apply to a three-variable system. In this way, all the permutations associated with Equation 5-14, specifically the cross-product terms, are displayed. For a three-variable system then, with nonindependent uncertainty sources, the uncertainty propagation equation is as follows in Equation 5-19. The result equation is Equation 5-18:

$$R = f(X,Y,Z) \tag{5-18}$$

$$
\begin{aligned}
(U_R)^2 &= (\partial R/\partial X)^2(U_X)^2 + (\partial R/\partial Y)^2(U_Y)^2 + (\partial R/\partial Z)^2(U_Z)^2 \\
&\quad + 2(\partial R/\partial X)(\partial R/\partial Y)U_{XY} + 2(\partial R/\partial X)(\partial R/\partial Z)U_{XZ} \\
&\quad + 2(\partial R/\partial Z)(\partial R/\partial Y)U_{ZY}
\end{aligned}
$$

$$
\begin{aligned}
&= (\partial R/\partial X)^2(U_X)^2 + (\partial R/\partial Y)^2(U_Y)^2 + (\partial R/\partial Z)^2(U_Z)^2 \\
&\quad + 2(\partial R/\partial X)(\partial R/\partial Y)b'_X b'_Y + 2(\partial R/\partial X)(\partial R/\partial Z)b'_X b'_Z \\
&\quad + 2(\partial R/\partial Z)(\partial R/\partial Y)b'_Z b'_Y
\end{aligned} \tag{5-19}
$$

where:

U_R =   a standard uncertainty in the result, either systematic or random

$U_X =$ a standard uncertainty in the measured parameter X, either systematic or random

$U_Y =$ a standard uncertainty in the measured parameter Y, either systematic or random

$U_Z =$ a standard uncertainty in the measured parameter Z, either systematic or random

$U_{XY} =$ the covariance of the X errors on Y errors

$U_{XZ} =$ the covariance of the X errors on Z errors

$U_{ZY} =$ the covariance of the Y errors on Z errors

$b'_X b'_Y =$ the product of the correlated systematic standard uncertainties in X and Y

$b'_X b'_Z =$ the product of the correlated systematic standard uncertainties in X and Z

$b'_Y b'_Z =$ the product of the correlated systematic standard uncertainties in Y and Z

It is obvious how complex this can get. However, simply being careful to follow the rules of mathematics will result in a proper uncertainty propagation. Note that, when the error sources are independent, all the correlation coefficients, ρ, are zero, as are all the covariances, so only the first three terms of Equation 5-19 remain for the independent case.

"Dithering" for Uncertainty Propagation

In the era of modern supercomputers, problems can be solved that have no closed-form equation as their solution. The closed-form equation permits the Taylor's Series approach to uncertainty propagation. Without that closed form, there is no way to perform a Taylor's Series uncertainty propagation.

Word of Advice about Dithering and the Need to Do It

Dithering is often not needed. In every case encountered by this author where there were extensive, detailed, highly complex, and lengthy expressions that required some iteration to solve, there was also in college texts a closed-form equation that would yield the correct result within a few percent of the expensive, long, and tiresome supercomputer method. While the long, tiresome computer methods are required for obtaining the most precise solutions (where, for example, an uncertainty in the equation of 2 or 3% is intolerable), the college text expression would yield the result to a few percent and an uncertainty analysis to likewise a few percent. That is, no one cares if the uncertainty in the uncertainty analysis is 10%. Many people do care if the value calculated for the result is in error by 10%. Therefore, continue your extensive computer simulations and overnight giga-mega bit runs for your test result, but try the college text simple expression for the basis of the uncertainty propagation. You'll likely find it works just fine.

"Dithering" Method

"Dithering" uncertainty propagation is a long process that proceeds approximately as follows. The computer software used to calculate the test or experimental result is used to calculate the result. Then, in turn, a convenient amount of each variable's standard uncertainty (systematic or random) is added to the input data to the software, and a new test result is calculated. The difference (between the new result and the old result) divided by the uncertainty increment used is the sensitivity of the result to uncertainty in that parameter at that result level. (The convenient amount of uncertainty added is usually 1%, 0.1%, 10°F, 1.0 psia, etc. It should approximate the expected systematic or random standard uncertainty, and its effect should be evaluated at or near each variable's level of interest as the slopes of equations can sometimes change significantly with a variable change of only a few percent.)

Note here that it is assumed that the uncertainty in the dependent axis parameter is symmetrical, as is the uncertainty in the independent axis parameter. If not, the methods of nonsymmetrical uncertainty estimation must be applied, sometimes to random uncertainties as well.

This process is repeated for all the uncertainty sources, and the resulting sensitivities are tabulated. The final uncertainty analysis is then computed by root-sum-squaring the systematic (and later the random) standard uncertainties times their sensitivities to obtain the result combined standard uncertainty, Equation 5-14.

Monte Carlo Simulation

This method of uncertainty propagation is even more difficult than dithering. Again, the computer software needed to obtain a test result is used. However, in this case, error distributions are posed for all uncertainty sources, systematic and random. A random number generator then selects an error from those distributions for each type of uncertainty source. Those errors are added to their corresponding parameter and the test result calculated. This process is repeated hundreds or thousands of times until a distribution of error in the test result is developed. The limits of this distribution are the uncertainty.

It is complicated, difficult, and almost always not needed. It does, however, provide a correct answer and can be easier than Taylor's Series approaches for very complicated equations.

5-4.   Uncertainty Propagation Examples

Simple Propagation Example, Independent Uncertainty Sources, Series Flowmeter Network

Given that the flow in a facility is measured by three turbine flowmeters in series, each of which has the same uncertainty, the best estimate of the facility flow is the average flow reported by the three meters. This condition is illustrated as follows:

Total flow $\rightarrow$ |meter A | $\rightarrow$ |meter B | $\rightarrow$ |meter C | $\rightarrow$ Total flow

That is:

$$F = (A + B + C)/3$$
$$= A/3 + B/3 + C/3 \tag{5-20}$$

where:

F = the average facility flow
A = the flow according to meter A
B = the flow according to meter B
C = the flow according to meter C

Utilizing Equation 5-10 for either systematic or random standard uncertainty, the following is obtained, assuming independent errors:

$$(U_F)^2 = [(\partial F/\partial A)^2(U_A)^2] + [(\partial F/\partial B)^2(U_B)^2] + [(\partial F/\partial C)^2(U_C)^2] \tag{5-21}$$

where U_X is the standard uncertainty in F, A, B, and C (systematic or random).

Substituting the partial derivatives, the following standard uncertainty equation is obtained:

$$(U_F)^2 = (U_A/3)^2 + (U_B/3)^2 + (U_C/3)^2 = (1/9)(U^2_A + U^2_B + U^2_C)$$

This is the simple case of linear addition models.

The case for parallel flowmeter networks is left as an exercise for the student (see Exercise 5-5); that is, what is the error propagation equation for three meters in parallel measuring the total flow?

Detailed Uncertainty Propagation Example, Some Uncertainties Correlated, Compressor Efficiency

Consider the example of the calculation of the measurement uncertainty of the efficiency measurement of a compressor. This task can be divided into several steps:

1. The defining equation for compressor efficiency

2. Making an uncertainty source summary

3. Defining the elemental pressure and temperature uncertainties

4. Propagating the uncertainties with Taylor's Series

5. Compiling an uncertainty analysis summary

These steps will now be handled in turn.

Step 1. The defining equation for compressor efficiency

The textbook definition of compressor efficiency is entirely adequate for this analysis. True, compressor designers have very detailed and fancy computer routines for this calculation, but that is not needed for a credible uncertainty analysis. The textbook expression will contain all the main factors that contribute uncertainty to the compressor efficiency calculation. That equation is:

$$\eta_c = [(P_2/P_1)^{(1-1/\gamma)} - 1]/[(T_2/T_1) - 1] \tag{5-22}$$

where:

η_c = compressor efficiency
P_1 = inlet pressure, nominally 14.7 psia
P_2 = exit pressure, nominally 96.0 psia
T_1 = inlet temperature, nominally 520°R
T_2 = exit temperature, nominally 970°R
γ = ratio of specific heats, nominally 1.40

(Note that the example levels for data in this example are the same as those in Ref. 5.)

Step 2. Making an uncertainty source summary

The next step in the uncertainty analysis is summarizing the uncertainty sources and their magnitudes. For simplicity, this example will detail the temperature measurement standard uncertainties and only outline the pressure measurement uncertainties.

The pressure and temperature measurement standard uncertainties may be summarized as shown in Table 5-1. These are the uncertainty estimates that apply to each measurement of each of the above parameters.

Note that in this uncertainty analysis the simplifying assumption has been made that all the random standard uncertainties are $s_{\bar{X}}$ and are obtained from enough data. Table 5-1 is only a summary of previous work. To fully understand the uncertainty analysis process, it is necessary to consider the elemental standard uncertainties and their magnitudes. This will be done for temperature only in this example, as shown in Table 5-2.

Table 5-1. Summarized Pressure and Temperature Measurement Standard Uncertainty Sources and Magnitudes

Parameter	Systematic (*b*)	Random ($s_{\bar{X}}$)	Degrees of Freedom
T1	0.41°R	0.60°R	32
T2	0.82°R	0.70°R	38
P1	0.010 psia	0.03 psia	45
P2	0.085 psia	0.17 psia	24

Note with care that the random standard uncertainty quoted is the standard deviation of the mean, $s_{\bar{X}}$, not just the standard deviation of the data, s_X. It is the standard deviation of the mean that is used in uncertainty propagation and in the Welch-Satterthwaite formula for combining degrees of freedom.

Step 3. Defining the elemental standard uncertainties

Tables 5-1 and 5-2 have both systematic and random standard uncertainties already defined. Note that the degrees of freedom are also noted for each of the random standard uncertainties. This is a vital part of the uncertainty analysis, for without those degrees of freedom, the proper *t*-statistic cannot be determined, and no uncertainty statement can be framed.

It is important to note that the exit temperature standard uncertainties have a portion of their systematic standard uncertainties that are dependent or correlated. These are the standard uncertainties that are identical to the standard uncertainties in the inlet temperature measurements because these exit and inlet temperature errors are also the same. Note that the same inlet temperature random standard uncertainties also show up in the exit temperature standard uncertainties summary but are not correlated with those of the inlet temperature. Why?

Table 5-2. Temperature Elemental Measurement Uncertainties

Uncertainty Source	Systematic (b)	Random ($s_{\bar{X}}$)	Degrees of Freedom (N − 1)	N+
Inlet temperature (T$_1$):				
TC* wire variability	0.25°R	0.20°R	45	1
Secondary TC reference	0.25°R	0.20°R	8	1
Ice reference	0.035°R	0.02°R	3	1
Data acquisition system	0.20°R	0.50°R	20	5
Root sum square	0.41°R(cr)	0.57°R	32**	
Exit temperature (T$_2$):				
All inlet errors	0.41°R(cr)	0.57°R	32	1
TC probe variability	0.50°R	0.30°R	16	1
TC drift	0.50°R	0.20°R	1	1
Root sum square	0.82°R	0.67°R	38 ++	

* TC = Thermocouple
**Actually 32.5, but truncating is proper for Welch-Satterthwaite degrees of freedom. Its calculation is left as an exercise for the reader.
+ N is the number of readings averaged at this hierarchy level. It is already in $s_{\bar{X}}$.
++ Actually 38.3, but truncating is proper for Welch-Satterthwaite degrees of freedom. Its calculation is left as an exercise for the reader.
(cr) These two uncertainties are identical; they are perfectly correlated (r = +1.00). This portion of the uncertainties must be treated specially in a Taylor's Series approach.

Random errors, although from the same error source, are never correlated or related. The fact that they have the same factor as their origin and the same magnitude does not make them correlated or dependent or related. The reason is that random errors cause a change in the test data every time they have a chance to do so. In addition, once done, the next opportunity has no relationship to the previous opportunity; random errors are random. Random uncertainties set bounds only on their distribution.

Systematic error, however, if it shows up more than once in an experiment, will do so at the same magnitude and sign each time. Here, some systematic error sources are the same and their error impacts are identical. After all, the definition for systematic error is that it does not change for the duration of the experiment or test.

It is for this reason that the 0.41°F systematic standard uncertainty in Table 5-2 is listed for both the root-sum-square of the inlet uncertainties and the "All inlet uncertainties" term of the exit temperature uncertainty summary. It is the estimate of the limits of the same, perfectly correlated, identical error. The correlation coefficient is +1.00.

In an uncertainty analysis, therefore, it is important to separate those uncertainties that are not dependent and those that are dependent, or correlated, or related. (Note here that correlation does not imply relationship, only correlation. This will be discussed later in Section 7-5.)

Throughout Tables 5-1 and 5-2, the uncertainty in γ, the ratio of specific heats, is considered negligible. This is the kind of assumption that makes the use of Taylor's Series for uncertainty propagation easy and yet sufficient for determining the measurement uncertainty with adequate correctness.

Step 4. Propagating Uncertainties with Taylor's Series

The next step in the uncertainty analysis is to propagate the pressure and temperature standard uncertainties into results units of efficiency.

To propagate the effects of the temperature and pressure standard uncertainties into efficiency units requires determining the partial derivatives of the efficiency equation with respect to four variables or parameters: inlet temperature, inlet pressure, exit temperature, and exit pressure. This can be done by dithering, by Monte Carlo simulation or, for this simple case, by partial differentiation.

Using Equation 5-14, the following sensitivities, or θ's, are obtained:

$$\theta_{T_1} = \frac{[T_2]\left\{[(P_2/P_1)^{[(\gamma-1)/\gamma]}]-1\right\}}{[(T_2-T_1)^2]} = +0.00340 \qquad (5\text{-}23)$$

$$\theta_{T_2} = \frac{[-T_1]\left\{[(P_2/P_1)^{[(\gamma-1)/\gamma]}]-1\right\}}{[(T_2-T_1)^2]} = -0.00182 \qquad (5\text{-}24)$$

$$\theta_{P_1} = \frac{-[(\gamma-1)/\gamma][(P_2/P_1)^{-1/\gamma}][P_2/(P_1)^2]}{[(T_2/T_1)-1]} = -0.03839 \qquad (5\text{-}25)$$

$$\theta_{P_2} = \frac{[(\gamma-1)/\gamma][1/P_1][(P_2/P_1)^{-1/\gamma}]}{[(T_2/T_1)-1]} = +0.00588 \qquad (5\text{-}26)$$

Step 5. Compiling an Uncertainty Analysis Summary

It is now convenient to summarize the uncertainty sources, their magnitudes, their parameters' nominal levels, and their sensitivities or influence coefficients. This is done in Table 5-3.

Table 5-3. Uncertainty and Source Summary

Error Source	Level (Units)	Systematic (b)	Random	Degrees of Freedom	Sensitivity
T_1	520 (°R)	0.41	0.57	32	+0.00340
T_2	970 (°R)	0.82	0.67	38	−0.00182
P_1	14.7 (psia)	0.010	0.03	45	−0.03839
P_2	96.0 (psia)	0.085	0.17	24	+0.00588

It is important to note that the systematic and random standard uncertainties are in the same units as the uncertainty sources, or parameters, *and that they are for one measurement of each parameter.* The sensitivities are in the units of [efficiency/(uncertainty source units)]. In this way, the sensitivities times the standard uncertainties will yield the effect of that uncertainty source in efficiency units. Those results may then be combined in root-sum-square methodology to yield the systematic and random uncertainty for the test or experiment.

Utilizing Equation 5-24, the following expressions for systematic and random standard uncertainties are obtained.

The systematic standard uncertainty one test result for efficiency in efficiency units is:

$$b_\eta = [(\theta_{T_1})^2(b_{T_1})^2 + (\theta_{T_2})^2(b_{T_2})^2 + (\theta_{P_1})^2(b_{P_1})^2$$
$$+ (\theta_{P_2})^2(b_{P_2})^2 + 2\theta_{T_1}\,\theta_{T_2}\,(b'_{T_1})(b'_{T_2})]^{1/2}$$
$$= [(0.00340)^2(0.41)^2 + (-0.00182)^2(0.82)^2$$
$$+ (-0.03839)^2(0.010)^2 + (0.00588)^2(0.085)^2$$
$$+ (2)(0.00340)(-0.00182)(0.41)(0.41)]^{1/2}$$
$$= [2.2273 \times 10^{-6} + 2.2273 \times 10^{-6}$$
$$+ 0.1474 \times 10^{-6} + 0.2498 + 10^{-6} - 2.0804 \times 10^{-6}]^{1/2}$$
$$b_\eta = [2.4873 \times 10^{-6}]^{1/2} = 0.0016 \text{ efficiency units} \tag{5-27}$$

Note that the prime (') indicates the use of only the correlated portion of the systematic standard uncertainties. That is, for this experiment, the

0.41°R is the same or identical for both the inlet and exit temperatures. For those systematic standard uncertainties, the correlation coefficient, r, is +1.00.

The expression for the random standard uncertainty for one test or experiment has exactly the same form:

$$s_\eta = [(\theta_{T_1})^2(s_{T_1})^2 + (\theta_{T_2})^2(s_{T_2})^2 + (\theta_{P_1})^2(s_{P_1})^2$$
$$+ (\theta_{P_2})^2(s_{P_2})^2]^{1/2}$$
$$= [(0.00340)^2(0.57)^2 + (-0.00182)^2(0.67)^2$$
$$+ (-0.03839)^2(0.03)^2 + (0.00588)^2(0.17)^2]^{1/2}$$
$$= [4.1616 \times 10^{-6} + 1.6231 \times 10^{-6}$$
$$+ 1.3264 \times 10^{-6} + 0.9992 \times 10^{-6}]^{1/2}$$
$$s_\eta = [7.1111 \times 10^{-6}]^{1/2} = 0.0027 \text{ efficiency units} \tag{5-28}$$

Note that the correlation, or covariance term (or cross-product term), is zero in Equation 5-28. This is because the correlation coefficient for the random uncertainties of the same source is always zero. They always have an independent effect on the uncertainty analysis. Once having occurred, the next opportunity does not know how the last one turned out. This is unlike the systematic terms of Equation 5-27.

The uncertainty still cannot be computed because the degrees of freedom associated with s_η as not been determined. This is necessary in order to select the Student's t from the table in Appendix D. The degrees of freedom are determined with the Welch-Satterthwaite formula:

$$\upsilon_\eta = \frac{[\sum(\theta_i s_i)^2 + \sum(\theta_i(b_i/2))^2]^2}{\left[\sum\left(\frac{(\theta_i s_i)^4}{\upsilon_i}\right) + \sum\left(\frac{(\theta_i(b_i/2))^4}{\upsilon_i}\right)\right]} \tag{5-29}$$

Using both the systematic and random uncertainties in Table 5-3, the following is obtained:

$$\upsilon_\eta = [(0.0034 * 0.60)^2 + ((-0.00182) * 0.70)^2$$
$$+ ((-0.03839) * 0.03)^2 + (0.00588 * 0.17)^2$$
$$+ (0.0034 * (0.41))^2 + ((-0.00182) * (0.80))^2$$
$$+ ((-0.03839) * (0.010))^2 + (0.00588 * (0.085))^2] \div$$

$$[(0.0034 * 0.60)^4/32 + ((-0.00182) * 0.070)^4/38$$
$$+ ((-0.03839) * 0.03)^4/45 + (0.00588 * 0.17)^4/24$$
$$+ (0.0034 * (0.41))^4/\infty + ((-0.00182) * (0.80))^4/\infty$$
$$+ ((-0.03839) * (0.010))^4/\infty + (0.00588 * (0.085))^4/\infty]$$

$$\upsilon > 30$$

The measurement uncertainty is then computed as follows:

$$U_{95} = \pm 2[(0.0016)^2 + (0.0027)^2]^{1/2}$$
$$= \pm 0.0063 \text{ efficiency units} \tag{5-30}$$

The calculation of uncertainty for the efficiency measurement is now complete. If it is desired to know the percent uncertainty, now is the time to calculate that. The efficiency level is 0.8198 ± 0.0063, or $0.8198 \pm 0.77\%$ for U_{95}.

When multiple measurements of a parameter are made (M times), the above s_i values would need to be divided by $\sqrt{M}$ before they are root sum-squared. This accounts for the averaging of several measurements of the parameter.

If the experiment is repeated M times, the s_η would be divided by $\sqrt{M}$ before it is used as the random uncertainty.

Suppose there were 20 T_1, 10 T_2, 15 P_1, and 5 P_2 measurements for each result and that the whole experiment was repeated three times ($M = 3$). The uncertainty would then be calculated as follows:

$$b_\eta = 0.0016 \text{ efficiency units (as before)}$$
$$s_\eta = [4.16160 \times 10^{-6}/\sqrt{20} + 1.62308 \times 10^{-6}/\sqrt{10}$$
$$+ 1.3264 \times 10^{-6}/\sqrt{15} + 0.99920\sqrt{20}/\sqrt{5}\ ]^{1/2}$$
$$= [(0.93056 + 0.51326 + 0.34248 + 0.44686) \times 10^{-6}]^{1/2}$$
$$= 0.0015 \text{ efficiency units} \tag{5-31}$$

$$U_{95} = \pm 2[(0.0016)^2 + (0.0015/\sqrt{3}\ )^2]^{1/2}$$
$$= \pm 0.0036 \text{ efficiency units}$$

The result uncertainty using Equation 5-31 is significantly lower than that obtained without the repeat testing and shown in Equation 5-30. Note that replicate measurements do not reduce the impact of systematic uncertainty at all.

Complex Uncertainty Propagation Example, Multiple Sources of Correlated Systematic Standard Uncertainties, Bolt Hole Diameter

Now let's determine the uncertainty in diameter, $\overline{D}$, of the bolt hole circle, A, B, C, D, in the plate illustrated in Figure 5-1.

Each bolt hole (no threads) diameter is measured once with micrometer $M1$. Each ligature length, AC and BD, is measured once with micrometer $M2$. (The ligature is the distance between the inside edges of opposing bolt holes.) Assume the systematic standard uncertainties of the two micrometers are independent of each other. What is the uncertainty of the bolt hole diameter, $\overline{D}$?

The equation for the diameter of the bolt hole is:

$$\overline{D} = (A/2 + B/2 + AC + C/2 + D/2 + BD)/2$$

where $A, B, C,$ and D are the bolt hole diameters of holes $A, B, C,$ and D, respectively, and where AC and BD are the ligatures of AC and BD, respectively.

Hint: Recall that for $P = F(X,Y)$,

$$U_P = \pm[(\partial P/\partial X)^2(U_X)^2 + (\partial P/\partial Y)^2(U_Y)^2$$
$$+ 2(\partial P/\partial X)(\partial P/\partial Y)r_{X,Y}(U_X)(U_Y)]^{1/2}$$

Assume more than 30 degrees of freedom in all cases and use the ASME/ISO U_{95} uncertainty model. Note that $r_{X,Y}(U_X)(U_Y)$ equals the covariance, $U_{X,Y}$.

$b_A, b_B, b_C,$ and b_D equal the systematic standard uncertainties of the diameters $A, B, C,$ and D. b_{AC} and b_{BD} equal the systematic standard uncertainties of the ligatures AC and BD. $s_A, s_B, s_C,$ and s_D equal the random standard uncertainties of diameters $A, B, C,$ and D. s_{AC} and s_{BD} equal the random uncertainties of ligatures AC and BD.

From uncertainty propagation, we obtain:

$$U_{\overline{D}} = [(\partial \overline{D}/\partial A)^2(U_A)^2 + (\partial \overline{D}/\partial B)^2(U_B)^2 + (\partial \overline{D}/\partial C)^2(U_C)^2$$
$$+ (\partial \overline{D}/\partial D)^2(U_D)^2 + (\partial \overline{D}/\partial AC)^2(U_{AC})^2$$
$$+ (\partial \overline{D}/\partial BD)^2(U_{BD})^2 + 2(\partial \overline{D}/\partial A)(\partial \overline{D}/\partial B)r_{A,B}(U_A)(U_B)$$
$$+ 14 \text{ more terms with all } r_{X,Y} \text{ combinations of } A, B, C, D, AC,$$
$$\text{and } BD]^{1/2}$$

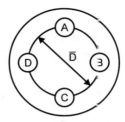

BOLT HOLE CIRCLE
A,B,C,D WITH
DIAMETER D

Figure 5-1. Bolt hole circle A, B, C, D with diameter $\bar{D}$

For the random standard uncertainties (all $s_{\bar{X}}$):

$$s_{\bar{D}} = [(\partial\bar{D}/\partial A)^2(s_A)^2 + (\partial\bar{D}/\partial B)^2(s_B)^2 + \partial\bar{D}/\partial C)^2(s_C)^2$$
$$+ (\partial\bar{D}/\partial D)^2(s_D)^2 + (\partial\bar{D}/\partial AC)^2(s_{AC})^2 + (\partial\bar{D}/\partial BD)^2(s_{BD})^2]^{1/2}$$

Note that $s_A = s_B = s_C = s_D$, that $s_{AC} = s_{BD}$, and that all $r_{X,Y} = 0$. Also,

$$\partial\bar{D}/\partial A = 1/4 = \partial\bar{D}/\partial B = \partial\bar{D}/\partial C = \partial\bar{D}/\partial D$$

and

$$\partial\bar{D}/\partial AC = 1/2 = \partial\bar{D}/\partial BD$$

Substituting, we obtain:

$$s_{\bar{D}} = [4(1/4)^2(s_A)^2 + 2(1/2)^2(s_{AC})^2]^{1/2}$$
$$= [((s_A)^2)/4 + ((s_{AC})^2)/2]^{1/2}$$

For the systematic standard uncertainties:

$r_{A,B} = r_{A,C} = r_{A,D} = r_{B,C} = r_{B,D} = r_{C,D} = 1$ (all same micrometer)

so their covariance terms are nonzero. That is,

$$u_{A,B},\ u_{A,C},\ u_{A,D},\ u_{B,C},\ u_{B,D},\ \text{and}\ u_{C,D} \neq 0$$

Also,

$r_{AC,BD} = 1$ (same micrometer, but
different from the one used on the holes)

so its covariance is nonzero. That is,

$$r_{AC,BD} \neq 0$$

All other $r_{X,Y} = 0$, that is:

$$r_{A,AC} = r_{A,BD} = r_{B,AC} = r_{B,BD} = r_{C,AC} = r_{C,BD} = r_{D,AC} = r_{D,BD} = 0$$

Their corresponding covariances are therefore zero. That is,

$$U_{A,AC} = U_{A,BD} = U_{B,AC} = U_{B,BD} = U_{C,AC} = U_{C,BD} = U_{D,AC} = U_{D,BD} = 0$$

Note that $U_{X,Y} = \sum U'_X(i)(U'_Y(i))$, the sum over i for M pairs of correlated systematic uncertainties. Here, the covariance equals the sum of the products of the pairs of correlated uncertainties in X and Y. For the example of the systematic uncertainties of A and AC, we have: $U_{A,AC} = b'_A b'_{AC}$, which equals $b_A b_{AC}$ in this example.

Substituting into the uncertainty propagation equation yields:

$$b_{\bar{D}} = [(\partial \bar{D}/\partial A)^2 (b_A)^2 + (\partial \bar{D}/\partial B)^2 (b_B)^2 + (\partial \bar{D}/\partial C)^2 (b_C)^2$$
$$+ (\partial \bar{D}/\partial D)^2 (b_D)^2 + (\partial \bar{D}/\partial AC)^2 (b_{AC})^2 + (\partial \bar{D}/\partial BD)^2 (b_{BD})^2$$
$$+ 2(\partial \bar{D}/\partial A)(\partial \bar{D}/\partial B)(b'_A)(b'_B)$$
$$+ \text{(6 more terms with all } b'_X b'_Y \text{ combinations of } A, B, C, D, AC,$$
$$\text{and } BD \text{ with covariances not equal to zero)}]^{1/2}$$

For this simple example, note that $b'_X = b_X$ and $b'_Y = b_Y$ throughout. Also, $b_A = b_B = b_C = b_D$ (same micrometer) and $b_{AC} = b_{BD}$ (same other micrometer).

Consequently,

$$b_{\bar{D}} = [4(1/4)^2 (b_A)^2 + 2(1/2)^2 (b_{AC})^2 + 6(2(1/4)(1/4)(1)b_A b_A)$$
$$+ 2(1/2)(1/2)(1)b_{AC} b_{AC}]^{1/2}$$
$$= [((b_A)^2)/4 + ((b_{AC})^2)/2 + (3/4)(b_A)^2 + (1/2)(b_{AC})^2]^{1/2}$$
$$= [(b_A)^2 + (b_{AC})^2]^{1/2}$$

Therefore, for the ISO/ASME U_{95} model (with more than 30 degrees of freedom in all cases), we have:

$$U_{95,\overline{D}} = \pm 2\{[(b_{\overline{D}}/2)^2] + [(s_{\overline{D}})^2]\}^{1/2}$$
$$= \pm 2\{[(b_A)^2 + (b_{AC})^2] + [((s_A)^2)/4 + ((s_{AC})^2)/2]\}^{1/2}$$
$$= \pm[(4(b_A)^2 + (b_{AC})^2) + ((s_A)^2) + (2(s_{AC})^2)]^{1/2} \ldots \text{q.e.d.}$$

Complex Uncertainty Propagation Problem and Answers to Key Questions, Measurement Uncertainty Analysis for Maintaining Constant Power Level

This particular problem is presented in detail with specific questions relating to its completion. Then the answers are presented in detail. We will determine the uncertainty in maintaining a constant power level for a power generation device given instrumentation uncertainties and operating control bands. (Note: control bands will need to be defined in statistical terms, not absolute limits.)

The operating equation is:

$$R = \frac{Q_{in}}{C_P} = \frac{mC_P(\Delta T)}{C_P} = m\delta T$$

where

R = power level
Q_{in} = heat input (BTU/hr)
ΔT = change in temperature across the test section (F)
C_P = specific heat at constant pressure (BTU/lbm-F), assumed = 1 (a constant)

Assuming all independent sources of uncertainty, we have the general expression:

$$(e_R)^2 = \left(\frac{\partial R}{\partial m}\right)^2 (e_m)^2 + \left(\frac{\partial R}{\partial \Delta T}\right)^2 (e_{\Delta T})^2$$

We now perform the partial differentiation and obtain:

$$\left(\frac{\partial R}{\partial m}\right) = \Delta T \text{ and } \left(\frac{\partial R}{\partial \Delta T}\right) = m$$

Both these expressions need to be evaluated at the test conditions of interest. The table following provides three test conditions of interest.

Power Test Conditions and Nominal Uncertainties

Line #	Parameter, level	Systematic Standard Uncertainty (b)	Random Standard Uncertainty ($s_{\bar{x}}$)
1.	For 16% power		
2.	Flow, @154 lb/hr		
3.	Orifice unc. (5%R)	7.7_B lb/hr	N/A*
4.	Control band	N/A	5_A lb/hr
5.	For 60% power		
6.	Flow, @ 236 lb/hr		
7.	Orifice unc. (5%R)	11.8_B lb/hr	N/A*
8.	Control band	N/A	5_A lb/hr
9.	For 100% power		
10.	Flow, @ 315 lb/hr		
11.	Orifice unc. (5%R)	15.8_B lb/hr	N/A*
12.	Control band	N/A	5_A lb/hr
13.	For 16%, 60% and 100% power		
14.	ΔTemperature, @ 20, 43 and 65F		
15.	TC calibration, each TC	$\langle 1.0_A \rangle$	N/A*
16.	ΔT unc. $(2)^{1/2}$	1.4_A F	N/A*
17.	T_{in} control band	N/A	$\langle 1_A$ F$\rangle$
18.	T_{out} control band	N/A	$\langle 1.5_A$ F$\rangle$
19.	ΔT control band $(2^2 + 3^2)^{1/2}$	N/A	1.8_A F

Note the subscripts "A" and "B" denote the ISO classifications of Type A and Type B uncertainties.

Assumption #1: In the above, we assume the B are 95% confidence uncertainties based on normally distributed errors, with infinite degrees of freedom and no correlation. These are equal to 2 standard uncertainties. The use of the capital letter "B" here is done for illustration purposes. Many documents do not use the lower case "b" when it can be assumed that all degrees of freedom are over 30. In those cases, it is sometimes easier to see and understand the magnitude of the systematic uncertainties when expressed as the 95% confidence interval instead of the 68% confidence interval of the lower case "b." Note too that when using this notation, that is capital "B," one needs to divide it by two (2) when it is included in the equation calculating uncertainty. That is done throughout this example.

Assumption #2: For the "control bands," we assume they are equivalent one standard deviation bands based on normally distributed errors.

Assumption #3: We assume no correlated error sources in any of the above.

The problem is to compute the uncertainty in power at each of the levels of interest and answer the following key questions.

1. What do the "$e's$" on page 137 represent? (Hint, they can represent many things as long as in an equation they all represent the same type of thing.)

2. Why is there an (*) by several "N/A's" in the table?

3. The uncertainties within carrots, $\langle \rangle$, are not to be included in the following uncertainty calculations. Why?

4. What would we do if assumption #1 was wrong and the B was based on only 5 degrees of freedom?

5. What would we do if assumption #2 was wrong and actually based on two standard deviations?

6. What would happen if assumption #3 was wrong? Which uncertainties would you think would likely arise from correlated errors?

7. How much different would the final uncertainty be if you grouped the elemental uncertainties by the ISO classifications, Type A and Type B, instead of systematic and random?

8. How would you estimate the uncertainty at 90% power? 40%?

Constant Power Level Uncertainty Answers

Computing the measurement uncertainty at 16%, 60% and 100% power (assuming no correlated errors):

$$U_{95} = 2\left[\left(\frac{B_R}{2}\right)^2 + (s_{\bar{X}, R})^2\right]^{1/2}$$

where:

$$b_R = \left[\sum_{i=1}^{N} \left(\frac{\partial R}{\partial P_i}\right)^2 (b_i)^2\right]^{1/2} \quad \text{and} \quad s_{\bar{X}, R} = \left[\sum_{i=1}^{N} \left(\frac{\partial R}{\partial P_i}\right)^2 (s_{\bar{X}, i})^2\right]^{1/2}$$

where P_i is the parameter associated with each b_i and $s_{\bar{X}, i}$. In this example, we have only two influence coefficients, or sensitivities: ΔT and m as shown in the problem statement.

Therefore, for 16% power, we have:

$$U_{95,\,16\%} = 2\left\{\left[(\Delta T)^2\left(\frac{7.7}{2}\right)^2 + (m)^2\left(\frac{1.4}{2}\right)^2\right] + [(\Delta T)^2(5)^2 + (m)^2(1.8)^2]\right\}^{1/2}$$

$$U_{95,\,16\%} = 2\left\{\left[(20)^2\left(\frac{7.7}{2}\right)^2 + (154)^2\left(\frac{1.4}{2}\right)^2\right] + [(20)^2(5)^2 + (154)^2(1.8)^2]\right\}^{1/2}$$

$$U_{95,\,16\%} = 2\{[(5929) + (11621)] + [(10000) + (76840)]\}^{1/2}$$

$$U_{95,\,16\%} = 646(R-units)$$

For 60% power, we have:

$$U_{95,\,60\%} = 2\left\{\left[(\Delta T)^2\left(\frac{11.8}{2}\right)^2 + (m)^2\left(\frac{1.4}{2}\right)^2\right] + [(\Delta T)^2(5)^2 + (m)^2(1.8)^2]\right\}^{1/2}$$

$$U_{95,\,60\%} = 2\left\{\left[(43)^2\left(\frac{11.8}{2}\right)^2 + (236)^2\left(\frac{1.4}{2}\right)^2\right] + [(43)^2(5)^2 + (236)^2(1.8)^2]\right\}^{1/2}$$

$$U_{95,\,60\%} = 2\{[(64364) + (27291)] + [(46225) + (180455)]\}^{1/2}$$

$$U_{95,\,60\%} = 1128(R-units)$$

For 100% power, we have:

$$U_{95,\,100\%} = 2\left\{\left[(\Delta T)^2\left(\frac{15.8}{2}\right)^2 + (m)^2\left(\frac{1.4}{2}\right)^2\right] + [(\Delta T)^2(5)^2 + (m)^2(1.8)^2]\right\}^{1/2}$$

$$U_{95,\,100\%} = 2\left\{\left[(65)^2\left(\frac{15.8}{2}\right)^2 + (315)^2\left(\frac{1.4}{2}\right)^2\right] + [(65)^2(5)^2 + (315)^2(1.8)^2]\right\}^{1/2}$$

$$U_{95,\,100\%} = 2\{[(263682) + (48620)] + [(105625) + (321489)]\}^{1/2}$$

$$U_{95,\,100\%} = 1720(R-units)$$

Below are the answers to the key questions.

1. The "e's" represent any uncertainty, b, $B/2$, s_X, $s_{\overline{X}}$, as long as the entire equation is consistent.

2. The asterisk (*) indicates that those uncertainty sources have no random component. They likely arise from calibration processes or standard practices.

3. Those uncertainties are portions of the line 19 uncertainty and can be combined since they are random and therefore from error sources that are uncorrelated.

4. We would need to divide the b terms by the appropriate Student's t that was used for those degrees of freedom to obtain the 95% confidence. That is, don't divide by 2 in the uncertainty equation, but by 2.57.

5. We would need to divide the control band uncertainties by 2 to obtain "1 standard deviation" for the uncertainty equation.

6. No difference at all. Identical results would result.

7. Plot a graph of the three uncertainties against their power levels and read the graph.

Now wasn't that fun?

5-5.   Detailed Equation for Uncertainty Propagation

To review and repeat in a format that is easily suitable for computer algorithms, note that the expression for combining random standard uncertainties with uncertainty propagation is:

$$(s_{\overline{X}, R})^2 = \sum_{i=1}^{N} [(\theta_i)^2 (s_{\overline{X}, i})^2] \tag{5-32}$$

where

$$\theta_i = \partial R / \partial V_i$$
$$V_i = \text{the } i\text{th measured parameter or variable}$$

The formula for combining systematic standard uncertainties with uncertainty propagation (Ref. 4) can be written as:

$$b_R = \left[\sum_{i=1}^{N} \left\{ [(\theta_i)^2 (b_i)^2] + \left[\sum_{j=1}^{N} \theta_i \theta_j b_{i,j}(1 - \delta_{ij}) \right] \right\} \right]^{1/2} \quad \text{(5-33)}$$

where the covariance term on the right, $b_{i,j}$, is the sum of the products of the correlated uncertainty sources and:

$$b_{i,j} = \sum_{l=1}^{M} b_i(\ell)b_j(\ell) \quad \text{(5-34)}$$

where

M = the number of correlated systematic uncertainties
ℓ = the ℓ th pair of correlated systematic uncertainties
δ_{ij} = 1 if $i = j$ and 0 if $i \neq j$ (the Kronecker delta)

Equation 5-33 is the general expression for how to handle correlated systematic uncertainties.

5-6.   Uncertainty of Results Derived from X-Y Plots

When a source of error appears in both the independent and the dependent axes of a data plot, the effect of that error source is to provide some measure of correlation or dependency of the errors in X on the errors in Y. Treating those uncertainties requires special handling.

A method for treating the uncertainties that arise from errors that appear in both X and Y has been developed by Price (Ref. 6) and is included in Ref. 7.

In brief, the treatment of these kinds of uncertainties is as follows.

Consider uncertainty sources that occur in an example such as a plot of the corrected fuel flow (CFF) versus the corrected shaft horsepower (CSH) of a gas turbine.

Random uncertainties or errors should be treated as independent by definition. Although a random uncertainty source may appear in both CFF and CSH, its occurrence is random and its impact on the X parameter does not affect its impact on the Y parameter (unless the readings are simultaneous, and then it too would be cause for an error that is in both variables and correlated). For this explanation, we'll assume that all random uncertainties are totally uncorrelated. We also note here the definition that a parameter is a value calculated from several variables. Here we have the parameters X and Y, or CFF and CSH.

So then, random uncertainties that arise from random errors which cause scatter about the line fit can be handled in uncertainty analysis directly as the scatter in the line fit. (More on this later in the section on curve fitting in Unit 7.)

Systematic uncertainties in Y propagate into the Y parameter directly; that is, they are also in the units of Y so their influence coefficients are unity. Systematic uncertainties in X propagate by the influence coefficient $\partial Y / \partial X$. However, a systematic error, i, that occurs in both X and Y is special, and its uncertainty must receive special treatment.

In the following discussion, it is assumed that the equation of the line fit is a straight line or, if it is a curve, it may be assumed to be a straight section that is tangent to the curve at the level of interest. In this way, the uncertainties are assumed to be small in comparison to the level of interest and symmetrical.

Figure 5-2 shows that an error in Y will move a line up or down. That is, a positive error in Y moves the line up and a negative error in Y moves the line down. This is to be expected.

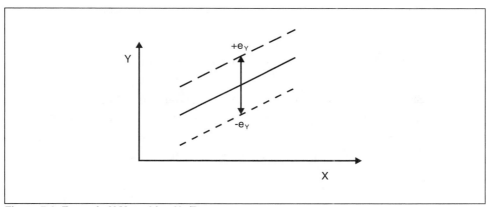

Figure 5-2. Errors in Y Move Line Up/Down

Figure 5-3 shows that an error in X will move a line right or left. That is, a positive error in X moves the line to the right and a negative error in X moves the line to the left. This, too, is to be expected.

Now consider the curious result shown in Figure 5-4. Here a positive error in Y is coupled with a positive error in X. The line first is displaced vertically by the positive error in Y and then is moved closer to its original position by the positive error in X. The other combinations of positive and negative errors in X and Y are also shown.

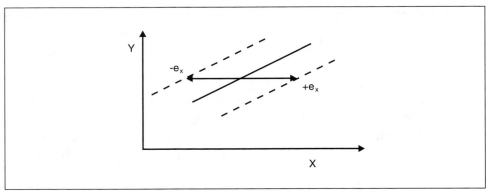

Figure 5-3. Errors in X Move Line Right/Left

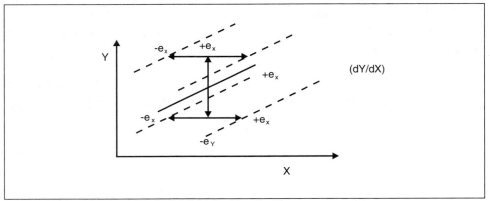

Figure 5-4. Errors in Both Y and X Affect the Line Position

In Figure 5-4, therefore, it can be seen that the net error is the difference between the Y error and the X error expressed in Y units (through the use of the sensitivity or influence coefficient). This is shown through the systematic uncertainty equations that follow.

The net uncertainty in Y can be expressed as:

$$b_{Y,net} = b_Y - b_X(\partial Y/\partial X) \tag{5-35}$$

where the symbols are obvious.

Note in Equation 5-35 that the minus sign allows for the fact that samesigned errors in X and Y have the effect of moving the resulting line closer to its original position.

Now for a particular uncertainty source, i, that is in both X and Y we have:

$$b_{Y,net,i} = b_{Y,i} - b_{X,i}(\partial Y/\partial X) \tag{5-36}$$

Equation 5-36 won't work because we need to accommodate the fact that the units of the systematic uncertainty, i, are likely not those of either X or Y. We must propagate that uncertainty source. The proper equation is:

$$b_{Y,net,i}= b_i(\partial Y/\partial i) - b_i(\partial X/\partial i)(\partial Y/\partial X) \tag{5-37}$$

The systematic uncertainties thus resulting need to be root-sum-squared with the other systematic uncertainties expressed in Y units.

Figure 5-5 shows the results of this type of analysis as presented by Price (Ref. 6). Note that were all the uncertainties considered to be independent, the most significant variable affecting the uncertainty of CFF and CSH would have been the inlet pressure, P_1. However, proper consideration of the uncertainties in both X and Y—that is, proper recognition that P_1 is in both CFF and CSH—results in the inlet pressure being the least important uncertainty source.

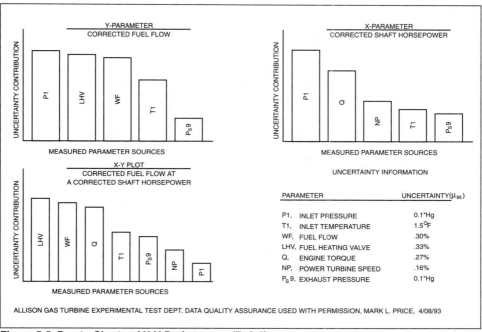

Figure 5-5. Pareto Charts of X-Y Performance (Ref. 6)

A detailed treatment of this approach was completed by Price, as noted in Ref. 6. The author heartily recommends this excellent paper to the interested reader. In addition, Ref. 1 has a much more detailed treatment of this problem.

5-7.   Summary

Uncertainty propagation, then, is required to evaluate the effect of uncertainty sources on a test result where either the uncertainty sources have different units or the result is in different units than the uncertainty sources. The easiest method for uncertainty propagation is to use the "college text" closed-form solution for the experimental test result and a Taylor's Series uncertainty propagation approach.

References

1. ANSI/ASME PTC 19.1-2006, *Instruments and Apparatus, Part 1, Test Uncertainty*, pp. 23–25.

2. Coleman, H. W., and Steele, W. G., Jr., 1989. *Experimentation and Uncertainty Analysis for Engineers*, pp. 189–199. New York: John Wiley & Sons.

3. Ku, Harry H., 1965. NBS Report No. 9011, "Notes on the Use of Propagation of Error Formulas." Washington, DC: National Bureau of Standards.

4. Brown, K. K.; Coleman, H. W.; Steele, W. G.; and Taylor, R. P., 1994. "Evaluation of Correlated Bias Approximations in Experimental Uncertainty Analysis." AIAA Paper 94-0772, presented at American Institute of Astronautics and Aeronautics 32nd Aerospace Sciences Meeting and Exhibit, Reno, NV.

5. ANSI/ASME PTC 19.1-1998.

6. Price, Mark L., 1993. "Uncertainty of Derived Results on X-Y Plots." In *Proceedings of the 39th International Instrumentation Symposium*. Research Triangle Park, NC: ISA.

7. ANSI/ASME PTC 19.1-1998, *Instruments and Apparatus, Test Uncertainty*.

Exercises:

5-1.   What is the purpose of uncertainty propagation?

5-2.   Orifice Problem

Given Test Pressure Data for an Orifice					
Point No.	P_{up}	P_{dn}	Point No.	P_{up}	P_{dn}
1	6.80	5.30	6	9.70	8.20
2	11.70	10.50	7	7.40	6.10
3	4.10	2.75	8	3.85	2.05
4	10.35	9.00	9	7.95	6.20
5	11.95	10.55	10	10.10	8.85

a. Compute the standard deviation of the upstream pressures, $s_{P_{up}}$, and the downstream pressures, $s_{P_{dn}}$.

b. Using $\Delta P = P_{up} - P_{dn}$

1. Obtain Taylor's Series for $s_{\Delta P_{ind}}$ (assume P_{up} and P_{dn} are independent).

2. Calculate $s_{\Delta P}$ with equations from 1) above (start with $s_{P_{up}}$ and $s_{P_{dn}}$).

3. Obtain Taylor's Series for $s_{\Delta P_{dep}}$ (assume P_{up} and P_{dn} are dependent).

4. Calculate $s_{\Delta P}$ with equations from 3) above (again use $s_{P_{up}}$ and $s_{P_{dn}}$). Note, $r = 0.9983$.

5. Calculate the 10 ΔP values.

6. Calculate $s'_{\Delta P}$ from the ΔP values. (The prime mearly indicates that calculated from the pressures themselves.)

7. Compare 6) with 4) and 2). What is learned?

5-3.   Word Problem

Aircraft in-flight net thrust equals engine gross thrust minus inlet drag. Measured airflow is a factor in calculating both drag and gross thrust. What problems does this produce in estimating the uncertainty in net thrust?

5-4.   Degrees of Freedom Problem

    a.   Calculate the degrees of freedom associated with the two root-sum-square random standard uncertainties in Table 5-2, 0.57°R and 0.67°R.

    b.   Calculate the degrees of freedom associated with the root sum square of the random standard uncertainties in Table 5-3.

5-5.   Parallel Flowmeters Uncertainty Propagation Problem

What is the uncertainty propagation equation for three meters in parallel measuring the total flow? That is:

$$\text{Total flow} \rightarrow \left| \begin{array}{c} \rightarrow \quad |\text{meter A}| \quad \rightarrow \\ \rightarrow \quad |\text{meter B}| \quad \rightarrow \\ \rightarrow \quad |\text{meter C}| \quad \rightarrow \end{array} \right| \rightarrow \text{Total flow}$$

5-6.   Decision Time Problem

    a.   Which is more accurate (lower uncertainty), total flow measured with three meters in series or with three meters in parallel? Assume all random standard uncertainties are ±1 gal/min and that all systematic standard uncertainties are independent and are ±1 gal/min.

    b.   Which method is more accurate if all the systematic errors (and uncertainties) are the same exact error source, that is, perfectly correlated?

5-7.   First Impression Problem

Parameter Q is obtained as follows:

$$Q = (T_1)^2 (T_2)$$

The typical systematic and random uncertainties are as follows:

	Level	b	$s_{\bar{X}}$	Units
T_1	25	3.0	3.0	°R
T_2	50	5.0	4.5	°R
Q	31250			

a. By inspection, note the largest and smallest uncertainty sources.

b. Compute the uncertainty of Q (assume $df > 30$).

The partial derivatives are:

$$\frac{\partial Q}{\partial T_1} = 2T_2(T_2)$$

$$\frac{\partial Q}{\partial T_2} = (T_1)^2$$

i. Calculate the systematic uncertainty component of the total uncertainty.
ii. Calculate the random uncertainty component of the total uncertainty.
iii. Calculate the total uncertainty of Q; use U_{95} in absolute units; in %.

c. Organize the data from the intermediate results of 2a and 2b above as follows:

$$\left(\frac{\partial Q}{\partial T_1}\right)^2 (b_{T_1})^2 \qquad \left(\frac{\partial Q}{\partial T_1}\right)^2 (2s_{T_1})^2$$

$$\left(\frac{\partial Q}{\partial T_2}\right)^2 (b_{T_2})^2 \qquad \left(\frac{\partial Q}{\partial T_2}\right)^2 (2s_{T_2})^2$$

Write out the values of the four above factors.

i. Which error source is largest? Smallest?
ii. How does this compare with your estimate in (1) above?
iii. What lesson is learned?

5-8. Relative Effect of Uncertainty Sources: Derivation Problem

a. For $N > 30$, derive the following equations which describe the fraction of the Total Uncertainty that b_i and s_i represent:

$$b_i \% TU = \frac{b_i^2}{(b_{RSS})(TU)}(100)$$

$$(s_i)\%TU = \frac{(s_i)^2}{(s_{RSS})(TU)}(100)$$

TU = Total Uncertainty

Hint: Relate fraction of the systematic term squared to (b_{RSS}) and that to its fraction of the Total Uncertainty (TU).

Repeat for (s_{RSS}).

b. Prove the total for 3 systematic and 3 random standard uncertainty terms is "1" (or 100%), i.e. $\Sigma[b_i \% TU] + \Sigma[(s_i) \% TU] = 1$.

Weighting Me
Multiple

UNIT 6

Weighting Method for Multiple Results

This unit addresses weighting several independent measurements in order to obtain a final test result that is more accurate than any of the individual measurements. It is a technique that is valuable where no one measurement method is adequate for the measurement uncertainty desired.

Learning Objectives—When you have completed this unit you should:

A. Understand the purpose of multiple test results.

B. Know the principles for calculating a final result that is weighted by the uncertainty of the several test results.

C. Be able to compute a final result by weighting, which is more accurate than any of the individual results used.

6-1. The Purpose of Multiple Measurements of the Same Test Result

It has been said that "the purpose for making redundant measurements is so the experimenter can choose the result he wants" (Ref. 1). Redundant, independent measurements of the same test result, when used properly, can yield a test result that is more accurate than any of the redundant measurements. This may be difficult to believe, but it is true. It is actually possible to average two measurements of the same results and get a final average result more accurate than either of the two. For example, if one measurement uncertainty is ±1.0% and the second is ±2.0%, the weighted result will have a lower uncertainty than ±1.0%. If both methods are ±1.0%, the weighted uncertainty will be at ±0.7%. This unit describes the weighting procedure, which will yield results with improved uncertainty when there is more than one measurement of the same result.

6-2. Fundamentals of Weighting by Uncertainty

A necessary condition for weighting results by their uncertainty is that the competing measurements be nondependent. A standard methodology for weighting competing results is found in many statistics texts (Ref. 2) and is covered in the national uncertainty standard (Ref. 3). The basic principles and approach will be covered in this unit.

Assumptions for this method are as follows:

A. More than one measurement of the same result is made by different methods.

B. The competing measurements and their errors are nondependent.

C. Each measurement method has its own, nondependent, uncertainty.

Fundamental Equations

There is a direct comparison between obtaining test results with weighted and without weighted averages. There are standard formulas for average, weighted average, systematic uncertainty, weighted systematic uncertainty, random uncertainty, weighted random uncertainty, degrees of freedom, and weighted degrees of freedom. These formulas will be presented with direct comparison to the usual formulas, which are unweighted.

Averages

The standard formula for an average of several measurements is:

$$\text{Grand Average} = \overline{\overline{X}} = \frac{\sum\limits_{i=1}^{N} N_i \overline{X}_i}{\sum\limits_{i=1}^{N} N_i} \qquad (6\text{-}1)$$

However, a better average—that is to say, an average with lower uncertainty—may be obtained from several independent (nondependent) measurements of the same test result, using a weighted average.

The formula for the weighted average, $\overline{X}_W$, is:

$$\text{Weighted Average} = \overline{X}_W = \sum\limits_{i=1}^{N} W_i \overline{X}_i \qquad (6\text{-}2)$$

where:

$$W_i = \frac{(1/U_i)^2}{\sum\limits_{i=1}^{N} [(1/U_i)^2]}$$

(6-3)

U_i is the uncertainty of the ith measurement method.

In this approach, the weighting is inversely proportional to the uncertainty of the method. That this is correct is intuitive. The most accurate single method, lowest uncertainty, should count the most. This approach does that.

For the case of two nondependent measurement methods yielding two nondependent test results, the situation is schematically as shown in Figure 6-1. Here, two different result averages can be obtained—one by the standard averaging technique, Equation 6-1, and the other by weighting, Equation 6-2. *When weighting is used, the resulting average will be known to* have greater accuracy (lower uncertainty) than even the best (lowest uncertainty) method employed.

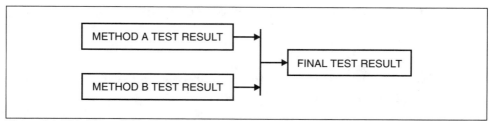

Figure 6-1. The Case of Two Non-Dependent Measurement Methods of a Single Test Result

For the simple case of only two measurement methods, the weights are:

$$W_1 = (1/U_1)^2/[(1/U_1)^2 + (1/U_2)^2]$$

(6-4)

$$= (U_2)^2/[(U_1)^2 + (U_2)^2]$$

$$W_2 = (U_1)^2/[(U_1)^2 + (U_2)^2]$$

(6-5)

Note that the sum of the weights W_1 and W_2 is unity.

$$W_1 + W_2 = 1.00$$

(6-6)

This is true for all weights for an experiment, no matter how many there are. It is a good check on the correct computation of weights or weighting factors.

Once the weighting factors are calculated, it is then possible to compute the uncertainty. Both weighted systematic uncertainty and weighted random uncertainty must be obtained and combined to obtain the weighted result uncertainty.

Weighted Systematic Standard Uncertainty Terms

Ref. 3 and 4 contain the equations used to obtain the weighted systematic standard uncertainty.

The usual calculation of systematic standard uncertainty is done with Equation 3-7:

$$b_R = [\Sigma(b_i)^2]^{1/2}$$

Equation 3-7 is used when all elemental systematic standard uncertainties, b_i, are the same units; they are in the units of the results. The influence coefficients are unity and implicit in the equation. The more general form of Equation 3-7 is Equation 6-7, where the influence coefficients, or sensitivity coefficients, θ_i, are shown explicitly:

$$b_R = [\Sigma(\theta_i)^2(b_i)^2]^{1/2} \tag{6-7}$$

Extending Equation 6-7 to the case of weighted systematic uncertainties, the following formula is obtained:

$$b_{R,W} = [\Sigma(W_i)^2(\theta_i)^2(b_i)^2]^{1/2} \tag{6-8}$$

Equation 6-8 is the expression that is used to compute the weighted systematic uncertainty to be associated with the weighted test result. The weighted systematic standard uncertainty is $b_{R,W}$.

Weighted Random Standard Uncertainty Terms

Ref. 3 and 4 contain the equations used to obtain the weighted random standard uncertainty.

The usual calculation of the random standard uncertainty is done with Equation 3-3:

$$s_{\overline{X}, R} = [\Sigma(s_{\overline{X}, i})^2]^{1/2}$$

Equation 3-3 is used when all random standard uncertainties, $s_{\overline{X}, i}$, are the same units; they are in the units of the results. The influence coefficients are unity and are implicit in the equation. The more general form of Equation 3-3 is Equation 6-9, where the influence coefficients, or sensitivity coefficients, θ_{ij}, are shown explicitly:

$$s_{\overline{X}, R} = \left[\Sigma(\theta_i)^2(s_{\overline{X}, i})^2\right]^{1/2} \tag{6-9}$$

Recall that for both Equations 3-3 and 6-9, Equation 3-2 applies. That is:

$$s_{\overline{X}, i} = [s_{\overline{X}, i}/(N_i)^{1/2}]$$

Extending Equation 6-9 to the case of weighted random standard uncertainties, the following formula is obtained:

$$s_{\overline{X}, R, W} = \left[\Sigma(W_i)^2(\theta_i)^2(s_{\overline{X}, i})^2\right]^{1/2} \tag{6-10}$$

Equation 6-10 is the expression that is used to compute the weighted random uncertainty to be associated with the weighted test result. The weighted random standard uncertainty is $s_{\overline{X}, R, W}$.

Weighted Total or Expanded Uncertainty

Once the weighted random uncertainty is obtained, it is only necessary to compute the weighted degrees of freedom in order to properly compute the weighted total uncertainty, U_W, if the uncertainty model is not the simplified U_{ASME} model. So, we will first consider the simpler case when we can assume that the U_{ASME} model applies. In that case, the weighted total or expanded uncertainty is computed as shown in Equation 6-11:

$$U_W = U_{ASME,W} = \pm 2[(b_{R,W})^2 + (s_{\overline{X}, R, W})^2]^{1/2} \tag{6-11}$$

Now for alternate confidences using Student's t of other than the 2.00 in Equation 6-11, we need to compute the weighted degrees of freedom.

Weighted Degrees of Freedom

The standard Welch-Satterthwaite formula for degrees of freedom assignable to a combined random uncertainties is Equation 3-4:

$$v = \frac{[\Sigma(s_i)^2]^2}{\{\Sigma[(s_i)^4/v_i]\}}$$

The weighted degrees of freedom is obtained by expanding Equation 3-4:

$$v_W = \frac{[\Sigma(W_i)^2(\theta_i)^2(s_{\overline{X},i})^2 + \Sigma(W_i)^2(\theta_i)^2(b_i)^2]^2}{\left\{\Sigma[((W_i)^4(\theta_i)^4(s_{\overline{X},i})^4)/v_i] + \Sigma[(W_i)^4(\theta_i)^4(b_i)^4/v_i]\right\}} \qquad (6\text{-}12)$$

Note that in Equation 6-12 the sensitivity coefficients, θ_i , are shown explicitly along with the weighting factors, W_i. Note also that the systematic uncertainties are included. The v_i are associated with their appropriate $s_{\overline{X},1}$ or b_i. The weighted degrees of freedom is v_W .

Uncertainty, Standard and Weighted

The standard uncertainty interval is given for U_{ASME} as:

$$U_{ASME} = \pm 2.00[(b_R)^2 + (s_{\overline{X},R})^2]^{1/2}$$

The corresponding weighted U_{95} uncertainty interval is:

$$U_{ASME,W} = \pm 2.00[(b_{R,W})^2 + (s_{\overline{X},R,W})^2]^{1/2} \qquad (6\text{-}13)$$

Equation 6-12 represents the weighted uncertainty that should be computed for any weighted test result.

If the assumption of Student's $t = 2.00$ is not valid or if an alternate confidence other than 95% is desired, the weighted uncertainty, $U_{95,W}$, is calculated as:

$$U_{95,W} = \pm t[(b_R)^2 + (s_{\overline{X},R})^2]^{1/2} \qquad (6\text{-}14)$$

Example 6-1:

For the simple case of two nondependent measurement methods, the following is assumed:

1. All influence coefficients are unity; the uncertainties are in result units.

2. The critical data for the two methods are as follows:

Method	Result	b_R	$s_{\overline{X}, R}$	v	U_{ASME}
A	5.0 psia	±0.1	±0.3	48	±0.63
B	7.0 psia	±0.05	±0.2	39	±0.41

Using Equation 6-1, the standard average is obtained:

$$\overline{X} = [(49)(5.0) + (40)(7.0)]/[49 + 40] = 5.90 \text{ psia}$$

Using Equation 6-2, the weighted average is obtained. First, however, the weights must be obtained. They are calculated with Equation 6-3:

$$W_A = (1/0.63)^2/[(1/0.63)^2 + (1/0.41)^2] = 0.298$$
$$W_B = (1/0.41)^2/[(1/0.63)^2 + (1/0.41)^2] = 0.702$$

Note that $W_A + W_B = 1.000$. This is a good check on the correct calculation of the weighting factors.

The weighted average can now be calculated using Equation 6-2:

$$\overline{X}_W = (0.298)(5.0) + (0.702)(7.0) = 6.4 \text{ psia}$$

This is the best answer or test result that can be obtained with the measurements given.

The weighted uncertainty can now be calculated. First, the weighted systematic standard uncertainty for the result (our average) is calculated with Equation 6-8:

$$b_{R,W} = [(0.298)^2(1.0)^2(0.1)^2 + (0.702)^2(1.0)^2(0.05)^2]^{1/2} = 0.046$$

The weighted random uncertainty can now be calculated with Equation 6-10:

$$s_{\overline{X}, R, W} = [(0.298)^2(1.0)^2(0.3)^2 + (0.702)^2(1.0)^2(0.2)^2]^{1/2} = 0.17$$

To calculate the weighted U_{ASME} uncertainty, Equation 6-13 is used:

$$U_{ASME,W} = \pm 2.00[(b_{R,W})^2 + (s_{\overline{X}, R, W})^2]^{1/2}$$
$$= \pm 2.00[(0.046)^2 + (0.17)^2]^{1/2}$$
$$= \pm 0.35$$

Note that $U_{ASME,W}$ is less than the uncertainty of the most accurate measurement method used, Method A, whose U_{95} was ± 0.41. The value of weighted uncertainties is now obvious.

If it cannot be assumed that there are enough degrees of freedom to set Student's t equal to 2.00, then we must determine the degrees of freedom with the Welch-Satterthwaite formula, Equation 6-15:

$$\upsilon_W = \frac{[\sum(W_i)^2(\theta_i)^2(s_{\overline{X}, i})^2 + \sum(W_i)^2(\theta_i)^2(b_i)^2]^2}{\left\{\sum[((W_i)^4(\theta_i)^4(s_{\overline{X}, i})^4)/\upsilon_{i1}] + \sum[(W_i)^4(\theta_i)^4(b_i)^4/\upsilon_i]\right\}} \quad (6\text{-}15)$$

The weighted degrees of freedom are then calculated as:

$$\begin{aligned}
\nu_W = &\{[(0.298)^2(1.0)^2(0.3)^2 + (0.702)^2(1.0)^2(0.2)^2 + \\
&(0.298)^2(1.0)^2(0.1)^2 + (0.702)^2(1.0)^2(0.05)^2]^2/ \\
&[(0.298)^4(1.0)^4(0.3)^4/48 + (0.702)^4(1.0)^4(0.2)^4/39 + \\
&(0.298)^4(1.0)^4(0.1)^4/\infty + (0.702)^4(1.0)^4(0.05)^4/\infty]\}
\end{aligned}$$

Remember that the degrees of freedom assumed for the elemental systematic uncertainties is infinity, ∞.

We then have:

$$\nu_W = \{(0.0013093)/(0.000011294)\} = 115.9 = 115$$

The weighted uncertainty is then calculated using Equation 6-14:

$$U_{95, W} = \pm t[(b_R)^2 + (s_{\overline{X}, R})^2]^{1/2}$$

In Appendix D we find that Student's t at 95% confidence is 2.00. The weighted uncertainty is then:

$$U_{95, W} = \pm(2.00)[(0.046)^2 + (0.17)^2]^{1/2} = \pm 0.35$$

Note that a weighted uncertainty of ±0.35 is better than the best single measurement uncertainty of ±0.41. Weighting works! However, in this example we see that $U_{95,\,W}$ is also a fine uncertainty estimate as it agrees with $U_{ASME,\,W}$.

References

1.   Dieck, R. H., 1987. "I made that one up myself."

2.   Brownlee, K. A., 1960. *Statistical Theory and Methodology in Science and Engineering*, pp. 72–74. New York: John Wiley & Sons.

3.   ANSI/ASME PTC 19.1-2006, *Instruments and Apparatus, Part 1, Test Uncertainty*.

Exercises:

6-1.   Why Weight Problem

    a. What is the primary reason for calculating a weighted result?

    b. What is the units requirement for a weighting computation?

6-2.   Weighting by Uncertainty Problem

Test Method	$\bar{X}$	b	$s_{\bar{X}}$	v
1	95°F	3°F	1.0°F	30
2	102°F	5°F	1.5°F	80
3	89°F	4°F	4.5°F	30

    a. Calculate the grand average for the three methods combined, $\bar{X}_G$. The weighting formula for W_1 is:

$$W_1 = \frac{(1/U_1)^2}{(1/U_1)^2 + (1/U_2)^2 + (1/U_3)^2} = \frac{(U_2\,U_3)^2}{(U_2 U_3)^2 + (U_1 U_3)^2 + (U_1 U_2)^2}$$

    b. Derive weighting formulas for W_2 and W_3.

    c. Evaluate values of W_1, W_2, and W_3. How can you check your results here?

    d. Compute: 1) weighted average, $\bar{X}_W$
                  2) weighted uncertainty, $U_{95,W}$

e. Compare the grand average with the weighted average ($\overline{X}_W$) and the weighted uncertainty ($U_{95,W}$) with the smallest (U_{95}) above. Why are they different?

Unit 7:
Applied Considerations

UNIT 7

Applied Considerations

This unit presents several data validation and analysis methods. These methods are valuable tools for an experimenter working on an uncertainty analysis. The methods are widely known and most are powerful. One or two are widely used but offer very misleading results for the unwary analyst. Often, measurement uncertainty standards and articles assume the experimenter already knows these techniques and how to apply them. Here, each will be presented so that an individual new to this type of analysis will be able to use them to improve the quality and applicability of an uncertainty analysis.

Learning Objectives—When you have completed this unit you should:

A. Understand how to select the proper units for errors and uncertainties.

B. Understand the basics of outlier rejection.

C. Understand the underlying assumptions and proper application of curve-fitting techniques.

D. Have a grasp of the use of and weaknesses associated with correlation coefficients.

E. Know how to test data for normality (probability plots).

F. Understand the basics of sampling error, its estimation, and its elimination.

7-1. General Considerations

This unit is a side trip from the path to a valid uncertainty analysis but an important one, since the use of these techniques will often reveal a problem in the test data that would negate the validity of the uncertainty analysis. Having these tools available will often keep the new as well as the experienced analyst out of trouble.

7-2. Choice of Error (and Uncertainty) Units

In previous units of this book, the subject of units has surfaced. It is important to have units that are consistent in every uncertainty analysis formula. In addition, the rule for choice of units is: *Choose error units that result in Gaussian-normal error data distributions.*

The reasons for this rule are: With Gaussian-normal error data, the following are well defined and well understood:

A. Average

B. Standard deviation

C. Variance (the square of the standard deviation)

D. Correlation

E. All other statistical terms used in uncertainty analysis

With anything but Gaussian-normal data, the above may not be well defined and their use could lead to unrealistic and/or erroneous conclusions.

In addition, as covered in Unit 5, matching units are needed for a viable and correct error or uncertainty propagation.

Method of Units Choice

The following example illustrates a method for proper units choice. Consider Figures 7-1 and 7-2, which are called *probability plots*. This is a special kind of data plotting covered in Section 7-6 in which Gaussian-normal data plot as a straight line. That is, if the data are normally distributed, they will appear as a straight line on this kind of paper.

It should be noted at this time that your local statistician probably has a computer program to both make these plots and to test the normality of data without plotting using a standard statistical test such as Bartlett's Test (Ref. 1). This plotting technique is recommended for most analysis efforts since the results can be seen and are thus more easily understood.

Returning to Figures 7-1 and 7-2, it is noted that the percent full-scale error data in Figure 7-1 are a straight line, indicating the errors are normally distributed. This means that the data are Gaussian-normal. These are the kind of data that should be used in an error analysis. Figure 7-2 illustrates the same exact error data but transformed into percent of reading. The data in Figure 7-2 are not normally distributed. These data are not to be used in uncertainty analysis.

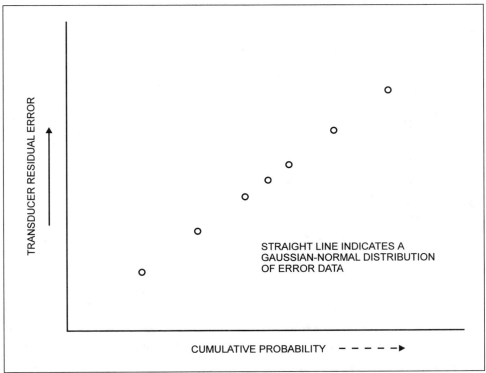

Figure 7-1. Probability Plot of Pressure Transducer Error Data, % Full-Scale Units

The only difference between the data plotted in Figure 7-1 and those in Figure 7-2 is a transformation of the same error data. It is easy to convert these data from one to the other, which is usually the case with other error data as well. For pressure transducers, usually the proper error data are in percent full-scale units, or absolute units, e.g., ±0.13 psi. Percent full-scale units are the same as absolute units divided by the full-scale absolute value.

Other instrument types are properly expressed as percent of reading units, such as turbine flowmeter error data and thermocouple error data.

Summary of Units

In summary, the proper units are needed for a correct uncertainty analysis. They are needed for error and uncertainty propagation and all the uncertainty analysis formulas. Several choices are available. The most common are percent of point or percent of level, percent full scale or absolute, or some combination. Care is needed to be sure that the units chosen result in the error data being Gaussian-normally distributed.

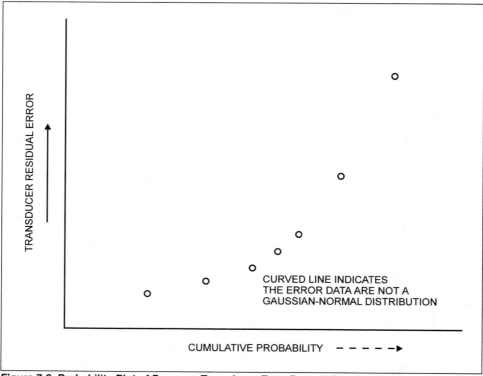

Figure 7-2. Probability Plot of Pressure Transducer Error Data, % Reading Units

7-3.   Treatment of Outliers

Definitions and Opinions

First of all, how is an outlier defined? Several definitions are available:

    A.   Outlier data are spurious data.

    B.   Outlier data are the same as "wild points" (whatever that means).

    C.   Outlier data are unexplained data.

    D.   Outlier data are data that disprove my thesis (this is my favorite definition).

This author eschews rejecting any data from a set as an outlier. There are two primary reasons for this position:

    A.   Outlier rejection is biased in the philosophical sense. That is, when did an experimenter ever reject a data point in order to better *disprove* the point being made? Think about it. Never, right? Outlier rejection is always used to better *prove* a point or to make a data set look better. It is biased toward the preconceived notions of the analyst, toward making a data set better prove the thesis or postulate. This kind of action should always be avoided if possible.

B.   The second reason this author *loathes* outlier rejection is that it is a poor way to improve data sets. If rejecting an outlier changes the conclusion of an experiment or set of measurements, a better set of data is needed. It is inappropriate for the conclusion(s) to rest on one data point.

Outlier Rejection Methods

Outlier data, therefore, should not be rejected. However, since it is human nature to make things look better, the following technique is given so that at least the approach to outlier rejection is objective. It is important to have an objective method for such rejection. There are two such common methods: Grubbs' Technique (Ref. 1) and Thompson's tau (τ) (Ref. 2).

Grubbs' Technique for outlier rejection is basically a 3σ technique. In essence, it rejects all data more than 3σ from the average. It seldom rejects good data (data within normal data scatter). Grubbs' Technique rejects few data points.

Thompson's tau (τ) is approximately a 2σ technique. It sometimes will reject good data (data within normal scatter).

The outlier rejection method recommended herein is Thompson's tau (τ).

Outlier Rejection Cautions

Several outlier rejection rules should be followed by the prudent analyst:

A.   Reject outliers only with caution.

B.   Never reject outliers by judgment (subjectively).

C.   Reject outliers only analytically (objectively).

Outlier Rejection Recommended Procedure: Thompson's tau (τ)

Consider the error data in Table 7-1. Note that, by observation, –555 may be an outlier. To determine whether or not it is, Thompson's τ technique should be applied as follows:

Step 1. Calculate the data set average, $\overline{X}$, and the standard deviation, s_X; note the number of data points, 40 in this case.

Here, $\overline{X} = 2.925$ and $s_X = 140.7$.

Step 2. Note the suspected outlier, –555.

Step 3. Calculate delta, $\delta = |(\text{suspect outlier}) - (\overline{X})|$, (absolute value).

Here, $\delta = |(-555) - (2.925)| = 557.925$.

Step 4. Obtain the Thompson's tau (τ) from the table in Appendix E.

Here, $\tau = 1.924$.

Step 5. Calculate the product of _ and S_x.

Here, $\tau s_X = (1.924) \times (140.7) = 270.7$.

Step 6. Compare δ with τs_X. If δ is $> \tau s_X$, X is an outlier. If δ is $< \tau s_X$, then X is not an outlier.

Here, $557.925 > 270.7$. Therefore, –555 is a candidate outlier.

Note that –555 is only a candidate outlier. This author does not recommend rejection without cause. Just passing the Thompson's τ test is not sufficient cause.

Table 7-1. Sample of Error Data

26	79	58	24	1	–103	–121	–220
–11	–137	120	124	129	–38	25	–60
148	–52	–216	12	–56	89	8	–29
–107	20	9	–40	0	2	10	166
126	–72	179	41	127	–35	334	–555

Step 7. Now search for the next most likely outlier. There are two approaches:

    a. In the strict interpretation of the Thompson's τ technique, one should return to step 2 above. Here, no new average or standard deviation is computed. This is the approach for automating the outlier detection method.

b. If however, it is desired to have a slightly more sensitive test, return to step 1. Then recompute the average, $\overline{X}$, and the standard deviation, s_X; follow through with steps 2 through 6.

Step 8. Whether step 7a or 7b is employed, the process is repeated until a suspected outlier yields a δ that is $< \tau s_X$. When the last X data point checked is not an outlier, the process stops.

In example calculations and in problem solutions in this book, the method using step 7b will be used.

Another Outlier Rejection Caution

This method for identifying outliers is to be used with caution. For the reasons cited previously, outlier rejection is to be done only with care. In addition, Thompson's τ will often identify valid data for rejection. For this reason as well, this outlier rejection method should be used to identify *possible* or *candidate* outliers for elimination from the data set should a good reason *also* be found. It is not advisable to eliminate outliers just to make the data set "look nice."

7-4.   Curve Fitting

General Comments

It has been said that "the use of polynomial curve fits is an admission of ignorance" (Ref. 3). Most experimenters have used curve-fitting routines to better present their data and to better understand the physical phenomena under investigation. However, the statistical assumptions underlying the use of curve-fitting algorithms now available are often ignored—if, in fact, the experimenter knows about them.

Curve-fitting routines available on modern computers are powerful tools that should be used with caution. They all have many underlying assumptions.

The foremost curve-fitting method available is that of least-squares (Ref. 4). It will be the method presented in this book.

Basic Assumptions of Least-Squares Curve Fitting

Least-squares curve-fitting methods have the following major statistically based assumptions:

A. The data being fitted has a continuous curve in the region of interest.

B. The curve being fitted has a continuous first derivative in the region of interest.

C. Over the range of interest, there is equal risk of error in the data.

The first two assumptions are "statisticalese,' "are very common in the world of the mathematician and statistician, and have little use for the uncertainty analyst. It is assumed that these first two assumptions will never be violated in experimental or test data sets.

The third assumption is often violated by scientists, engineers, and experimenters. It means that when the curve is fit, there is equal risk of error in the data around the curve. That is, the data scatter looks the same across the whole range of the curve in the region of interest. Figures 7-3 and 7-4 illustrate this point.

Figure 7-3 shows a data set in which the scatter around the curve fit is the same across the whole range of the curve. This is equal risk of residual error. It is the result to which all curve-fitting efforts should aspire.

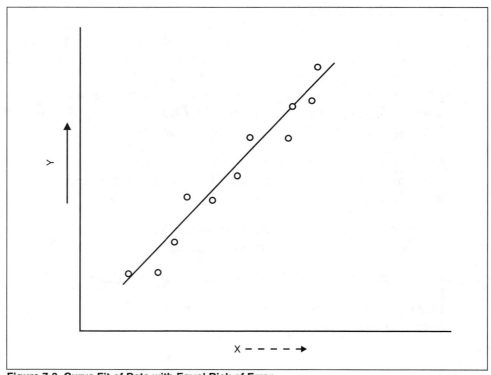

Figure 7-3. Curve Fit of Data with Equal Risk of Error

Figure 7-4 shows a curve fit in which the data scatter is larger at the lower end than at the higher end of the curve. This is *not* equal risk of error. The basic underlying assumptions of statistics have been violated and the curve fit should not be used. This curve fit is typical of the plot of nozzle calibrations. At high flows and Reynolds numbers, the data scatter is usually small, a testimony to the low random uncertainty of the pressure transducers that measure the delta pressure across the nozzle. At low flows and Reynolds numbers, there is considerable scatter in excess of that at the high end. This, too, is typical of the results obtained with typical pressure transducers on a nozzle. This kind of curve fit is anathema. The competent analyst will never use it.

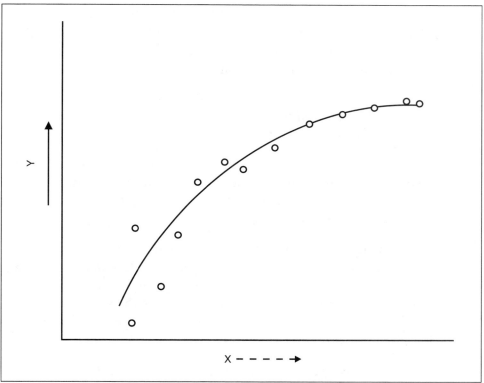

Figure 7-4. Curve Fit of Data without Equal Risk of Error

A note of interest: When curve fits of the type in Figure 7-4 show up, they can often be corrected by a transformation of the data, possibly to semilog. However, these transformations can be very tricky, so watch out!

Additional Guidance and Cautions

In addition to the fundamental statistical assumptions behind valid curve fitting, here are a number of helpful hints this author has learned to heed over his many years of mistakes.

    A.   *Avoid the use of polynomial curve fitting wherever possible.*

As mentioned before, its use is an admission of ignorance. Note: *"Stupid is forever, ignorance can be fixed"* (Ref. 5). Of what is the experimenter ignorant when polynomial curve fits are used? (He is not stupid, certainly, just ignorant.) The experimenter is ignorant of the physics of the problem. The physical form is not known. The theoretical relationship is not known.

This may be stated another way: *Use the theoretical relationship wherever possible.* It will keep the experimenter out of trouble. The most likely spot for this trouble is the space between the data points. This fact is also related to the number of data points fitted and the type of curve being fit.

    B.   *The number of constants calculated in a curve fit should be much less than the number of data points* (Ref. 6).

Every curve fit computes a number of constants or coefficients. These constants may be for the physical form or for the polynomial. The number of constants should be much less than the number of data points or vast trouble will surely be encountered between the data points when one attempts to use the curve fit.

When the number of constants calculated equals the number of data points, the curve fit will go exactly through each data point, giving the appearance of a fine fit. However, to go through those points, gigantic gyrations will occur between the points. Since the primary use for any curve fit is to extract information between the points, this can be devastating to any serious analysis.

More on the real problems of violating hints A and B later.

    C.   *2SEE for curve fits is analogous to $2s_X$ for data.*

$2s_X$ for data is an expression and description of the scatter of the data around an average which is displaced by systematic error from the true value. (Remember that all averages are displaced from the true value, and an important part of any uncertainty analysis is to estimate the limits of that systematic error, the systematic uncertainty. The effects of systematic errors cannot be seen in the data as the effects of random errors can.)

SEE stands for *standard estimate of error*. It is a description and estimation of the scatter of the data about a fitted line or curve. The formulas for s_X and SEE are comparable. Note Equations 2-3 and 7-1:

$$s_X = \left[\frac{\sum_{i=1}^{N}(X_i - \overline{X})^2}{N-1} \right]^{\frac{1}{2}} \tag{2-3}$$

$$SEE = \{[\Sigma(Y_i - Y_{ci})^2/(N-K)]^{1/2}\} \tag{7-1}$$

where for both:

N = the number of data points in the s_X or the fit
K = the number of constants in the fit
$\overline{X}$ = the data set average
X_i = the ith data point in the data set or fit
Y_{ci} = the Y calculated at X_i from the curve fit
Y_i = the ith measurement of Y in the fit
$N-1$ = the degrees of freedom associated with s_X
$N-K$ = the degrees of freedom for the curve fit

Note that both equations have exactly the same form. Both s_X and SEE are expressions of the scatter of the data. s_X is scatter about the average, and SEE is scatter about the fitted curve or line.

> D. *A low SEE is not necessarily an indicator of a good curve fit. Corollary: For multiple polynomial curve fits, the one with the lowest SEE is not necessarily the best curve fit to use.*

A low $2s_X$ is indicative of a low random error in the experiment. It is often thought that a low $2SEE$ is indicative of an accurate (or the best) curve fit. One can get into trouble with that assumption, as can be seen in Table 7-2.

Table 7-2. Comparison of 2SEE Obtained for Several CO2 Curve Fit Orders

Fit Order	2SEE in %CO_2
1	0.194
2	0.052
3	0.058
4	0.039
5	0.007

When observing the 2SEE values in Table 7-2, it could be assumed that the lowest 2*SEE* (for 30+ d.f.) would be indicative of the best curve fit, the one that should be used by the experimenter. Few things could be farther from the truth. The lowest 2*SEE* is for curve fit order 5; it is actually the worst curve fit! (This will be illustrated in Figure 7-5.) Its use also ignores hint A: Use theoretical relationships wherever possible and avoid polynomial curve fits wherever possible.

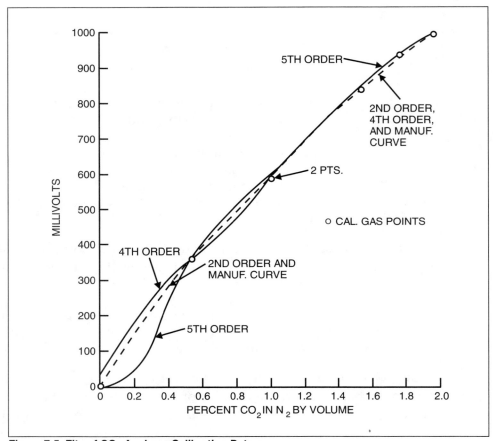

Figure 7-5. Fits of CO$_2$ Analyzer Calibration Data

E.   *The 2SEE of a curve fit should approximate the error in the standards* used in the calibration.

It is not reasonable to have less data scatter about a curve fit than the comparative scatter in the calibration standards. This is illustrated in the next section.

Example of Curve-Fitting Problems

The problems covered in hints A through D are all illustrated in Figure 7-5, which presents CO_2 calibration data and its several curve fits to show that polynomial curve fits can cause problems, that the number of constants calculated should be less than the number of data points, that *SEE* is analogous to s_X, and that the lowest 2*SEE* is not necessarily an indicator of the best curve fit.

Figure 7-5 shows several curve fits. Polynomial curve fits for the second, third, fourth and fifth order are shown for CO_2 analyzer calibration data consisting of seven data points. Also shown is the manufacturer's recommended curve form, an exponential.

Note that in Figure 7-5 the second-, third-, fourth-, and fifth-order curve fits that are illustrated are the curve fits whose 2*SEE* values are listed in Table 7-2. It is apparent how badly the fifth-order curve fits the proper shape for this exponential instrument. Particularly between 0.0 and 0.6% CO_2, substantial between-point errors are encountered when using the curve. This is the reason a curve fit should have many fewer constants calculated than the number of data points available for fitting.

The next question is: "What is the 'number of fit constants' that is 'much less' than the number of data points?" First realize that "fit constants" applies to least squares curve fits. What is intended by the question is: "What is a number of constants that is 'much less' than the number of data points in the fit?"

Observe in Figure 7-5 that there are seven data points. Obviously, a fifth order fit with six constants is not "much less" than the number of data points. How about fourth order (five constants)? That is also no good. There are still substantial between-point errors encountered when using the curve. However, second order appears to be quite good. In fact, it overlays the manufacturer's exponential curve. Second order has three constants, fewer than half the number of data points. This would seem a good start. *The curve fit should have at least twice the number of data points as the number of constants calculated in the fit.* In the case of the second-order fit, another curve fit was also tried, an exponential of the form:

$$Y = ae^{bX} \tag{7-2}$$

This exponential form has only two constants and was the best fit since it overlaid the manufacturer's curve and the simple second-order mentioned before. This simple fit, with two constants, did so well because it is actually the physical form. It conforms with hint A: Use the physical form whenever possible.

Note also that using the lowest $2SEE$ value in Table 7-2 as an indicator of the best curve fit does not work. The best fit was actually second order. Its $2SEE$ was 0.052% CO_2. This brings us to the impact of hint F: The $2SEE$ of a curve fit should approximate the uncertainty in the standards used in the calibration. Consider the data in Table 7-3. Review the data and consider which curve fits are proper and which are "overfits" of the calibration data. This will be left as an exercise for the student.

Table 7-3. CO_2 Instrument Polynomial Calibration Curves

Instrument	Range	Fit Order	%CO_2 2SEE	(Standard) Cal. Gas Error Uncertainty %CO_2
CO_2—A	0–5%	3	0.028	0.044
	0–2%	3	0.020	0.022
CO_2—B	0–18%	6	0.101	0.115
	0–5%	6	0.035	0.044
	0–2%	5	0.0073	0.022

Least-Squares Curve Fitting

Since it is important to use curve fits that match the physical form wherever possible, it is important to know how to do a least-squares fit on an equation. Most least-squares curve fits are done on polynomials, and the illustrations in this unit will be on polynomials. However, it should be clearly noted that these methods will apply to other forms as well, as for example, a simple exponential given as Equation 7-2.

This section presents the basics of least-squares curve fitting. The fit illustrated will be that of a straight line. Consider Figure 7-6. The form for a straight line is:

$$Y = aX + b \qquad (7\text{-}3)$$

Note in Figure 7-6 that epsilon (ε) is also shown. ε is used to represent the error between the line fit and the data points:

$$\varepsilon = Y - (aX + b) \qquad (7\text{-}4)$$

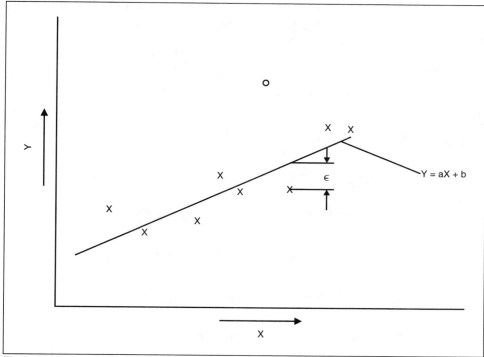

Figure 7-6. Least-Squares Line Fit

To derive the least-squares equations for the straight-line fit, the following process is used.

The line form is:

$$Y = aX + b$$

For one point on the line, the following is true:

$$Y_i = aX_i + b \qquad\qquad (7\text{-}5)$$

For one point off the line, the following is true:

$$Y_i = aX_i + b + \varepsilon_{Yi} \qquad\qquad (7\text{-}6)$$

The term ε_{Yi} represents the error between the point (X_i, Y_i) and the Y_i value of the line fit at X_i. To obtain a least-squares fit, the sum of the squares of these ε_{Yi}'s must be minimized; that is, it is necessary to minimize $\Sigma\,[(\varepsilon_{Yi})^2]$.

It is noted that:

$$\Sigma\,[(\varepsilon_{Yi})^2] = \Sigma\,[(Y_i - aX_i - b)^2] \qquad\qquad (7\text{-}7)$$

For $\Sigma\,[(\varepsilon_{Yi})^2]$ to be minimized, Equation 7-7 must be differentiated with respect to a and b and those differentials set equal to zero. a and b are the floating constants for this curve fit.

Differentiating the square of Equation 7-7 and setting the results equal to zero, the following are obtained:

$$[\partial/\partial a][\textstyle\sum(Y_i^2 + a^2 X_i^2 + b^2 - 2Y_i a X_i - 2Y_i b + 2ab X_i)] \; = \; 0.0 \qquad (7\text{-}8)$$

$$[\partial/\partial b][\textstyle\sum(Y_i^2 + a^2 X_i^2 + b^2 - 2Y_i a X_i - 2Y_i b + 2ab X_i)] \; = \; 0.0 \qquad (7\text{-}9)$$

Dropping noninvolved terms, the following are obtained:

$$[\partial/\partial a][\textstyle\sum(a^2 X_i^2) - 2\sum(Y_i a X_i) + 2\sum(ab X_i)] \; = \; 0.0 \qquad (7\text{-}10)$$

$$[\partial/\partial b][\textstyle\sum(b^2) - 2\sum(Y_i b) + 2\sum(ab X_i)] \; = \; 0.0 \qquad (7\text{-}11)$$

The resulting simultaneous equations are:

$$0 \; = \; 2a\textstyle\sum X_i^2 - 2\sum Y_i X_i + 2b\sum X_i \qquad (7\text{-}12)$$

$$\begin{aligned}
0 \; &= \; 2\textstyle\sum b - 2\sum Y_i + 2a\sum X_i \\
&= \; 2Nb - 2\textstyle\sum Y_i + 2a\sum X_i
\end{aligned} \qquad (7\text{-}13)$$

Working through the algebra, the solutions for a and b are obtained.

From Equation 7-12:

$$b \; = \; [\textstyle\sum(Y_i X_i) - a\sum(X_i)^2]/[\sum(X_i)] \qquad (7\text{-}14)$$

By substituting into Equation 7-13 and rearranging, the solution for the constant a is obtained:

$$a \; = \; \left\{[N\textstyle\sum(Y_i X_i) - \sum(X_i)\sum(Y_i)]/[N\sum(X_i)^2 - (\sum(X_i))^2]\right\} \quad (7\text{-}15)$$

Substituting Equation 7-15 into Equation 7-14 and reorganizing, the solution for constant b is obtained:

$$b \; = \; \textstyle\sum(Y_i)/N - a\sum(X_i)/N \; = \; \bar{Y} - a\bar{X} \qquad (7\text{-}16)$$

The application of these principles is covered in the exercises.

7-5. Correlation Coefficients: Their Use and Misuse

General Comments

Probably one of the most misunderstood and misused statistics is the correlation coefficient. It is needed for uncertainty analysis when errors are not completely independent, as covered in Section 5-3.

However, although an indicator, the correlation coefficient is not proof of cause and effect. This is the point on which most experimenters and novice uncertainty analysts find themselves frequently impaled. *Correlation is not necessarily a demonstration of cause and effect.* This fact is hard to believe for some but, as proof, consider the following situation.

If one were to calculate a correlation coefficient between the price of haircuts in New York City and the population of Puerto Rico for the past 20 years, a strong correlation would be obtained; that is, the correlation coefficient would be near one. Since correlation coefficients can be only in the range of −1.0 to +1.0, one would think that getting a correlation coefficient near +1.0 would indicate relationship and/or cause and effect. However, as can be intuitively seen, there is no cause and effect between the population of Puerto Rico and the price of haircuts in New York City. There is no cause and effect.

The Correlation Coefficient

The correlation coefficient for an infinite data set, or the whole population, is ρ; for a data sample, it is r. In uncertainty analysis, as with the calculation of s_X or the choice of σ, one seldom has the whole population and must calculate r, not ρ, just as one calculates s_X and not σ. The formula for the correlation coefficient for the data sample is, therefore:

$$r_{XY} = \frac{N\sum(Y_iX_i) - (\sum X_i)(\sum Y_i)}{\left\{[N\sum(X_i^2) - (\sum X_i)^2][N\sum(Y_i^2) - (\sum Y_i)^2]\right\}} \tag{7-17}$$

Equation 7-17 looks like an easy visit to your calculator, but it is not. It is worth calculating r once for training purposes; after that, do it with a programmable calculator or a computer.

Correlation Coefficient as an Estimator of Explained Variance

The correlation coefficient squared, r^2, is often used as an estimation of the variance in one error source that is explained by the second. That is, Equation 7-18 can be used in uncertainty analysis to decide how much variance in one variable, $(s_X)^2$, may be explained by variance in the second correlated error source variance, $(s_Y)^2$.

$$(r_{XY})^2 = \frac{\text{Explained variation}}{\text{Total variation}} \tag{7-18}$$

Although Equation 7-18 is true, it is not often needed in uncertainty analysis. It is useful at times, however.

The Significant Correlation Coefficient: Analytical Approach

Although the graphical test for the significance of r is visual, and intuitive, a strict analytical approach is also useful. Both Pearson and Hartley (Ref. 7) and Hald (Ref. 8) have shown an expression using the correlation coefficient that follows the Student's t distribution:

$$t = \frac{|r|\sqrt{N-2}}{\sqrt{1-r^2}} \tag{7-19}$$

where

t	=	Student's t at a confidence of choice
r	=	the sample correlation coefficient
$(N-2)$	=	the degrees of freedom for the correlation coefficient
N	=	the number of pairs of data points used to compute r

Equation 7-19 is used to calculate a Student's t from a sample correlation coefficient, r. If the absolute value of the Student's t calculated is not greater than the absolute value of the two-sided Student's t distribution for that number of degrees of freedom and for the confidence of interest, then the data sets have not been shown to be correlated with that confidence. In Hald's words, "If the t value corresponding to a given value of r is smaller than, say the 90% fractile, the data in hand do not justify the conclusion that the two variables are correlated" (Ref. 8).

Then, *also very important here*, the Taylor's Series uncertainty propagation formulas need not employ the terms with the correlation coefficients that do not pass this test.

Remember too that p, the population correlation coefficient, is estimated by r, the sample correlation coefficient, that is:

$$r \approx p \qquad\qquad (7\text{-}20)$$

7-6. Tests for Normality, Probability Plots

General Comments

In all the combinations of standard deviation mentioned in this book, there has been the implicit assumption that the data sampled are normally distributed. It is appropriate to test for this normality.

A quantitative test for normality is the Kolmogorov-Smirnov Test Statistic (Ref. 9). This and other tests provide fine quantitative answers to the question of normality.

However, it is often more instructive to the uncertainty analyst to use a more qualitative technique called probability plotting (Ref. 9). A detailed explanation of probability plotting is given in Ref. 10 and is covered using clear illustrations in J. R. King's book (Ref. 11).

Besides the clear indication of normality, much can be learned from a probability plot of the error data, as will now be seen.

Probability Plotting Procedure

The procedure for making probability plots involves the following steps:

Step 1. Arrange the data in an increasing list, lowest value to highest.

Step 2. Assign "median ranks" to each numbered data point using Benard's formula (Ref. 11) following:

$$P_{0.50} = (i - 0.3)/(N + 0.4) \qquad\qquad (7\text{-}21)$$

where:
$P_{0.50}$ = the 50% median rank
i = the number of the ordered data point being plotted
N = the number of data points plotted

Step 3. Plot the data against its median rank (or cumulative percent, which is the median rank times 100) on normal probability paper. One type of normal probability paper that is available is Keuffel & Esser Company's No. 46-8003.

Step 4. Observe the shape of the curve plotted and decide if it is sufficiently close to a straight line to be considered normally distributed data. Data that form a straight line on this special paper are normally distributed.

Alternate Method for Probability Plotting (without Special Paper)

If the special paper is not available, an identical plot of the data can be made by altering step 3 above as follows:

> Step 3. After assigning median ranks, determine from the table in Appendix F the equivalent fraction of sigma, σ, associated with that median rank. (Interpolate the table in Appendix F linearly for adequate precision.) Then plot the data set, in order, against its sigma fraction, σ, on rectilinear graph paper. The resulting curve shapes will be identical to those obtained with the special probability paper.

It is good to note that this sigma fraction can be obtained with a common spreadsheet software package such as Microsoft Excel. The statistical formula to use is "NORMSINV." This returns the plus and minus sigma fraction associated with the median ranks calculated and expressed as fractions (not multiplied by 100 for the special plotting paper).

Then proceed with step 4 as shown.

Additional Information in Probability Plots

Probability plots offer additional information besides an assessment of data normality:

A. The slope of the straight line is proportional to the standard deviation of the data.

B. The 50% point is the average of the data set.

To determine the data set standard deviation, find the 50% and 2.5% points on the Y axis, subtract them, and divide by 2; this difference is the s_X interval.

To determine data set average, just read the vertical axis at the point where the straight line of the data goes through the 50% point (or the 0.0 σ point if plotting against σ has been done because probability plotting paper was not been available).

Examples of Probability Plots

When data are plotted on probability paper, the shapes of the curves are often most informative. When they are straight lines, the data are normally distributed. When there are curve shapes other than a straight line, the shapes themselves are revealing.

Consider Figures 7-7(a) and (b). Figure 7-7(a) is the typical plot of a normal distribution. When such a data set is plotted on probability paper, Figure 7-7(b) results. It is a straight line, indicating the data set is normally distributed.

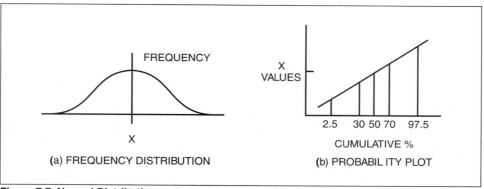

Figure 7-7. Normal Distribution

Now consider Figures 7-8(a) and (b). Figure 7-8(a) is a typical plot of a data set in which two normal distributions are included in the set, but their averages are different and there is no overlap of the data. This is often called a bimodal distribution. Note: In Figure 7-8(b) is the probability plot of the data in Figure 7-8(a); there are two separate straight lines. This indicates two distributions in one data set. It might be thought that Figure 7-8(b) is unnecessary since one can see the bimodality in Figure 7-8(a). However, a real data set is not so neatly separated as shown in Figure 7-8(a). Figure 7-8(b) separates it neatly. Here, the bimodular data have been separated into two straight lines, each with its own average and standard deviation.

Now consider the data in Figures 7-9(a) and (b). Here the bimodular data exist in two distributions that overlap. When this happens, as in Figure 7-9(a), the probability plot will look like the data in Figure 7-9(b), that is, an "S" shape.

Look at the data plotted in Figures 7-10(a) and (b). Two frequency distributions of data are plotted in Figure 7-10(a). They are not bimodular,

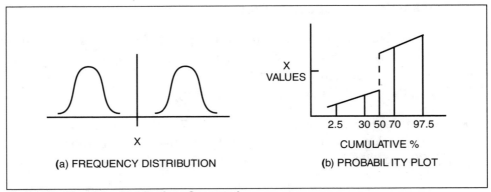

Figure 7-8. Bimodular Distributions Separated

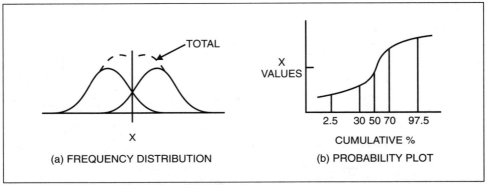

Figure 7-9. Bimodular Distributions Overlapped

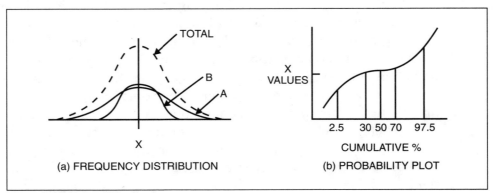

Figure 7-10. Nested Distributions

but nested. The probability plot of Figure 7-10(b) shows this by having a reverse "S" curve. It is the inverse "S" as compared to Figure 7-9(b) for bimodular data.

Finally, consider the data in Figure 7-11. Actual differences obtained by comparing two CO_2 instruments on the same process stream have been

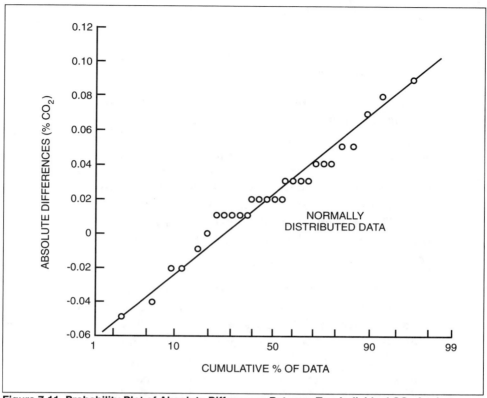

Figure 7-11. Probability Plot of Absolute Differences Between Two Individual CO$_2$ Analyzers

probability plotted. At first it appears to be normally distributed; that is, if s_X were calculated, it could be properly used because the data are normally distributed.

However, note also that the data are stacked up every 0.01% CO$_2$. This is called a "least-count" problem; it indicates that the instruments used to compute these delta CO$_2$ readings could not yield data with any finer increments than 0.01% CO$_2$. This is not usually a problem in uncertainty analysis except to note that the problem exists and that it would be unreasonable to try to measure a difference smaller than 0.01% CO$_2$.

7-7. Sampling Error

Special Considerations

Sampling error is often misunderstood by the novice uncertainty analyst. It can be used to represent any of three (or more) error types or sources. The most common three are:

A. Variation in space

B. Variation in time

C. Sampling error

The most common uncertainty analysis problem is when *real variation is misunderstood to be random error.* This real variation can be real variation in space or time. True sampling error is *the residual random error after accounting for all the real effects, such as those of space and/or time.*

One way to evaluate the effects of space and/or time is to see your local statistician and have him or her run an analysis of variance on your test data. The problems with this approach are that most experimenters don't understand what is happening and that results are evident only after completion of the analysis.

This section presents a simplified analysis of variance technique sometimes called "matrix analysis." This matrix analysis technique will be first presented in all the gory detail of the full analytical equations. It will then be applied to a simple example so that the student may more clearly understand what is happening. This technique assumes there are no interactions present between time and space.

Gory Detail Analytical Equations, Method

1. Given:

    A. Data (measurements) at a set of spatial locations such as in a pipe or a duct.

    B. Several data acquisitions of the data at each location but spaced in time. It is assumed that all the locations are sampled at the same time but all are sampled several times.

2. The operating equation is defined as follows:

    $$\text{Data acquired} = D_{lt} \tag{7-22}$$

    where:

    $$
    \begin{aligned}
    D &= \text{the actual data point measurement} \\
    l &= \text{the location of the measurement in the duct} \\
    t &= \text{the time of the measurement (this is the time at} \\
      &\quad \text{which all the data are recorded at all locations)}
    \end{aligned}
    $$

3. The next step is to obtain the average at each time across all locations. This is defined by Equation 7-23:

$$A_t = \sum_{l=1}^{N} (D_{lt})/N \qquad\qquad (7\text{-}23)$$

where:

A_t = the average of all data at time, t, across all locations, l

N = the number of locations, l

4. It can now be observed, by considering the averages and their times, whether or not there is an effect of change in time. This is something that cannot be seen during an analysis of variance, but which can be seen here.

5. Next, obtain the differences (δ) by comparing the data at each location to the average at that time, that is:

$$\delta_{lt} = D_{lt} - A_t \qquad\qquad (7\text{-}24)$$

where δ_{lt} = the difference between the data at each location, l, and its time, t, average.

6. Now it is necessary to obtain the average of the differences, δ_{lt}, at each location across time, that is:

$$\bar{\delta}_1 = \sum_{t=1}^{M} (\delta_{lt})/M \qquad\qquad (7\text{-}25)$$

where:

$\bar{\delta}_1$ = the average of all δ_{lt} at location, l, across time, t

M = the number of times averaged

7. Next it is necessary to obtain the differences, Δ_{lt}, comparing each time difference, δ_{lt}, to its average at location, l, as shown in Equation 7-26:

$$\Delta_{lt} = \delta_{lt} - \bar{\delta}_1 \qquad\qquad (7\text{-}26)$$

8. Note that the Δ_{lt} values are the residual errors after the linear variations in time and space are averaged out.

9. The next step is to calculate the sampling error, s_X, using Equation 7-27:

$$s_X = \left\{\left[\sum\sum(\Delta_{lt} - \bar{\Delta})^2\right]/[(M-1)(N-1)]\right\}^{1/2} \tag{7-27}$$

where $\bar{\Delta} = (\sum\sum\Delta_{lt})/(M \times N)$.

10. It is important to note that the sampling error thus calculated, s_X, should be treated as a random error source and used as a random uncertainty.

It is much easier to understand the calculation of sampling error using the matrix analysis approach when the calculation is presented in real data. Therefore, use the previous section to define the equations needed for the calculation and load them into your computer. In order to understand them, however, consider the following analysis.

Example 7-1:

1. Consider the data or measurements that could be obtained with five thermocouples measuring the temperature in a pipe or full duct, as shown in Figure 7-12.

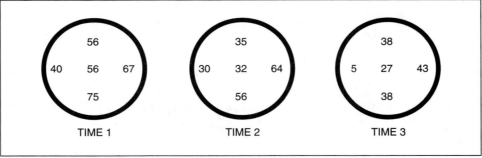

Figure 7-12. Data for Five Location Temperatures and Three Times in a Duct

2. The temperature data can be reorganized in a data matrix of test temperatures, °F, as shown in Table 7-4. Note that the position number need not be rationally assigned to an actual position.

3. Obtain the average at each time, as shown in Table 7-5.

Table 7-4. Pipe Temperature Data Matrix, Location and Time

Times	Locations				
	1	2	3	4	5
1	56	56	75	40	67
2	35	32	56	30	64
3	38	27	38	5	43

Table 7-5. Pipe Temperature Data Matrix With Time Averages

Times	Locations					Ave. (A_t)
	1	2	3	4	5	
1	56	56	75	40	67	58.8
2	35	32	56	30	64	43.4
3	38	27	38	5	43	30.2

4.   Note that it is possible at this point to observe an effect of time; that is, for later times, the temperature of the fluid in the pipe is lower.

5.   Now obtain the differences by comparing each location value to the average at its time.

6.   Then obtain the average of those differences at each location across time, as shown in Table 7-6.

7.   Obtain the differences by comparing each time, δ_{lt}, to its average at location 1, $\bar{\delta}_{lt}$, to obtain the residual errors in Table 7-7.

8.   The values in Table 7-7 are the residual errors after the linear effects of time and location are removed by this "matrix" analysis.

9. $s_X = 6.8°F$. The effect on the average for the pipe is $s_{\bar{X}} = 6.8/\sqrt{15}$, which equals $1.8°F$.

10. s_X should be treated as the random uncertainty resulting from another random error source. This also assumes that there is no "interaction" among these two variables, time and location.

Note that the standard deviation of the original data was $s_X = 18.2°F$. There has been a considerable reduction in apparent sampling error through this analysis.

Table 7-6. Pipe Temperature Data Matrix Differences from Time Averages with Location and Time Averages

Times	Locations					Ave. (A_t)
	1	2	3	4	5	
1	−2.8	−2.8	16.2	−18.8	8.2	58.8
2	−8.4	−11.4	12.6	−13.4	20.6	43.4
3	7.8	−3.2	7.8	−25.2	12.8	30.2
Ave. (δ_{lt})	−1.1	−5.8	12.2	−19.1	13.9	

Table 7-7. Pipe Temperature Data Matrix with Residual Errors After Removing Linear Effects of Time and Location

Times	Locations				
	1	2	3	4	5
1	−1.7	3.0	4.0	0.3	−5.7
2	−7.3	−5.6	0.4	5.7	6.7
3	8.9	2.6	_4.4	_6.1	_1.1

In the foregoing, it would have been incorrect to have used the variation observed in the pipe as an estimate of the sampling error. The standard deviation of ±18.2°F was unrealistically high and included the real effects of location and time. The correct answer, or standard deviation or random uncertainty, is the ±6.8°F just calculated.

7-8. Selecting Instrumentation Based on Vendor Uncertainty Statements

One would think that with all the current technology available, instrumentation vendors would make use of it to better describe the capability of their instruments to perspective customers. In fact, what happens is each vendor seeks to present his device in the best light possible. This is no surprise. However, each therefore provides uncertainty (or accuracy) information in their own unique format and includes only their individually selected metrics of uncertainty, such as "accuracy," "repeatability," "zero drift," "linearity" and the like. The following several tables illustrate this situation. All are summaries from a catalogue of instrumentation. The information is taken directly from the vendor quotes or statements in that catalogue.

Pressure Transducer Instrumentation Accuracy Statements

The following is a table of statements made by five **pressure transducer companies**.

Vendor	Accuracy	Hysteresis	Non-Linearity	Sensitivity Drift	Zero Balance
A	±0.5% F.S.	—	—	—	±1% F.S
B	±1%	—	—	<2%	—
C	±0.5%	±0.01%	±0.05%	±1.5%	—
D	±10%	±25%	—	—	—
E	—	—	—	—	—

How should one use the above table of comparison data to select the proper vendor for a pressure transducer? It cannot be done! Note that some numbers are given in %F.S. and some in just %. How do you compare them? Some data is not given, especially for vendor "E." Oh, and how vendor "C" can claim ±0.5% accuracy when he has an admitted ±1.5% sensitivity drift is beyond my comprehension.

Now consider the following table of comparison data for **temperature instruments**.

Vendor	Accuracy	Linearity
Bimetalic	±1%	—
Pyrometer	Yes	—
RTD A	Greater	—
RTD B	±0.1%	<1%

Note three of four say nothing about linearity. Note especially the accuracy quote for the Pyrometer vendor, "Yes." I guess he means, "Yes, we have accuracy." How much, we don't know. Note too the first RTD vendor, he has "Greater" accuracy. Greater than what? We don't know. How can we make an informed selection here? We cannot.

Now consider the following table of **gas analysis instruments'** performance statements.

Here too, comparisons are impossible. Not all vendors give all metrics. For Repeatability and Zero Drift, many of the vendors use different units.

Vendor	Accuracy	Repeatability	Sensitivity	Zero Drift	Span Drift
A	—	±1%F.S.	—	±1% F.S./week	—
B	—	±2%	—	—	—
C	—	—	±0.4	—	—
D	±2%	±0.5%F.S.	—	±1%F.S./day	<2%F.S./day
E	—	±1%	—	<±1%/day	<±1%/day

Finally, consider this table of performance metrics for **flowmeters**.

Vendor	Accuracy	Repeatability	Sensitivity	Linearity
A	—	±1%	±0.1%	—
B	±2%R	±0.1%F.S.	—	±0.5%F.S.
C	±0.5%R	±0.2%F.S.	—	—
D	—	±0.1%	—	±0.1%

Once again, the units change with the vendor and sometimes within a single vendor.

In all the above, no clear definitions of the terms used, such as accuracy and repeatability, is given either. It looks like there was no attempt to conform to any standard for measurement uncertainty estimation. We are left with little or no information from which to make an informed choice for new or improved instrumentation.

The Solution to Evaluating Vendor Uncertainty Statements

This writer recommends seeking out vendors that will allow you to evaluate their instrumentation in your working environment. That way, you can decide if its performance is suitable to your needs. It requires work, but yields answers useful to the individual application.

References:

1.   Grubbs, F. E., 1969. "Procedures for Detecting Outlying Observations in Samples." *Technometrics* 11, 1:00–00.

2.   Thompson, W. R., 1935. "On Criteria for the Rejection of Observations and the Distribution of the Ratio of the Deviations to Sample Standard Deviation." *Annals of Mathematical Statistics* 6:214–219.

3. Dieck, R. H., 1987. "I made that one up."

4. Mace, C., 1967. *Essentials of Statistics for Scientists and Technologists*, pp. 106–115. New York: Plenum Press.

5. Dieck, R. H., c. 1981. "A sign in my office."

6. Dieck, R. H., c. 1975. "One of my neat little rules."

7. Pearson, E. S., and Hartley, H. O., Editors, 1962. *Biometrika Tables for Statisticians*, Vol. I, Cambridge, U.K.: University Press. pp. 28–30, 87–89, 140–141.

8. Hald, A., 1960. *Statistical Theory with Engineering Applications*, pp. 608–609. New York: John Wiley & Sons.

9. Bowker, A. H., and Lieberman, G. J., 1972. *Engineering Statistics*, 2nd ed., pp. 454–458. Englewood Cliffs, NJ: Prentice Hall.

10. Ibid., pp. 452–454.

11. King, J. R., 1971. *Probability Charts for Decision Making*, pp. 52–65. New York: Industrial Press.

Exercises:

7-1.   Outliers: Thompson's tau, τ, Problem

Consider this data set:

1	−103	−121	−220
129	−38	25	−60
−56	89	8	−29
40	2	10	166
127	−35	334	−555

a. For given data without point −555, determine whether or not 334 is an outlier.

b. If it was, what is the next point to check? For increased sensitivity, here, re-calculate $\bar{x}$ and s_X. Is this point an outlier? If so, go to "c."

c. Are there any more outliers?

7-2.   Curve-Fitting Problem (see Table 7-3)

    a. What is the curve fit with the lowest $2SEE$ for instruments CO_2—A and B?

    b. Are those lowest $2SEE$'s indicative of the best fits for the instruments and ranges noted?

    c. What instruments and ranges have properly sized $2SEE$'s when the calibration gas errors are considered? Why?

7-3.   Least-Squares Curve-Fit Problem:

Derive the equations needed for a least-squares curve fit that is second order and forced through zero. (Hint: Use $Ax^2 + Bx + C = y$, where $C = 0$.) Remember to minimize [sum of squares of error term ε]: $\varepsilon = Ax^2 + Bx - y$.

7-4.   Correlation Problems

Test Data Sets					
A		**B**		**C**	
X	Y	X	Y	X	Y
0.8	0.5	1.7	−1.9	0.7	1.0
−0.7	−1.6	−7.8	−6.2	−2.0	3.2
4.8	4.3	−1.9	1.6	−1.5	0.8
2.6	2.4	2.8	1.5	2.2	−.0
6.2	6.0	1.1	−0.9	1.8	−0.5
3.5	3.1	9.5	7.0	−0.3	0.2
−0.8	−0.9	1.4	2.7	−3.3	1.8
1.8	2.2	−2.6	−0.3	2.1	−1.4
−2.4	−2.3	2.5	−1.0	−1.2	−1.7
−0.2	0.4	−1.1	−1.9	3.4	−1.2
0.9	1.2	0.7	1.8	2.0	1.6
3.5	3.7	−0.8	1.7	−2.5	2.2

The correlation coefficients r_A, r_B, and r_C are +0.986, +0.795, and −0.628, respectively. Using Equation 7-19:

    a. Is r_A significant at 95%?

    b. Is r_B significant at 95%?

    c. Is r_C significant at 95%?

    d. Plot X versus Y for each data set.

e. What is noticed about the plots? Recommend action.

f. Compute a new r_B. Is it significant at 95%?

g. What is learned?

7-5.   Probability Plot Problem

Data Sets: Error Data or Test Data			% Median Ranks	
Set A	Set B	Set C	15 Points	12 Points
−0.25	0.10	3.60	5	6
−0.85	0.57	1.40	11	14
0.20	−0.42	−1.10	18	22
−0.65	0.76	2.90	24	30
−2.70	−1.80	4.20	31	38
−1.30	0.38	−2.80	37	46
0.85	1.08	3.20	44	54
−1.70	0.25	3.90	50	62
−0.50	−0.23	0.00	56	70
−0.05	−0.05	0.70	63	78
1.40		3.40	69	86
−2.10		2.10	76	94
−1.00			82	
0.45			89	
−1.50			95	

a. Assign median ranks to sets A and C. Remember to order the data from lowest to highest first (largest negative number is lowest). Calculate and assign median ranks to data set B.

b. Make up the probability plots for data sets A, B, and C. Remember to multiply the median ranks by 100 to get cumulative %. Median ranks for data sets "A" and "C" are already done in the Table above.

c. What is observed for each:
    1. Line straight? Conclusion?
    2. $\overline{X}$?
    3. s_X?
    4. Outliers?

d. Calculate $\overline{X}$ and s_X for data sets A, B, and C. Compare with those in "c" above.

e. Any action recommended?

Unit 8:
Presentation of Results

UNIT 8

Presentation of Results

A frequently neglected part of any uncertainty analysis is the formulation of a compelling presentation of the content and impact of the results. The conclusion of an uncertainty analysis should yield information that a manager or experimentalist finds invaluable for the purposes of decision making. The primary purpose of doing an uncertainty analysis is to provide information that describes just how much faith can be placed in the results of a test. In this unit, the presentation of the results of a measurement uncertainty analysis will be detailed.

Learning Objectives—When you have completed this unit you should:

A. **Understand the need for a forceful, concise presentation of the results of an uncertainty analysis.**

B. **Be knowledgeable concerning the several presentation formats available.**

C. **Know how to assemble a compelling summary of uncertainty analysis results for management and customers.**

8-1.   General Considerations

The presentation of the results of an uncertainty analysis must be orderly, complete, and compelling.

It must be *orderly* so that a novice (read: manager) can understand and apply the information.

It must be *complete* so that the competency of the analyst is never called into question. Most who use the results of an uncertainty analysis don't understand its origin and just barely understand how to apply it. If there is an error or omission in the presentation, it will serve only to undermine the credibility of the analyst and the analysis. Credibility to an analytical person is all-important.

The presentation must be *compelling* so that its impact will be felt and its conclusions applied to the test results.

8-2.   Presentation Content

The presentation of the uncertainty analysis results should include the following:

    A.   The total uncertainty, either U_{95} or U_{ASME}
- systematic standard uncertainty
- random standard uncertainty
- degrees of freedom
- uncertainty model chosen (from above two choices)

    B.   The elemental standard uncertainties
- in measurement units
- in result units

    C.   Relative effect of uncertainty sources
- in percent of nominal level
- in percent of total uncertainty

    D.   Summary of the uncertainty propagation

8-3.   Illustration

The simplest way to present a complicated uncertainty analysis is to illustrate it. Here the illustration is for the uncertainty analysis of calculated turbine efficiency. It will be noted throughout that the detailed, boring, long, and gory equations that competent turbine performance analysts use for turbine efficiency are missing. Only five uncertainty sources are noted, although there are really dozens. Only the main drivers are presented here.

The uncertainty analysis presentation should start with the bottom line. Don't waste time and lose credibility by trying to give a manager too much detail at first. Work into it as needed for all to whom these results will be presented. Two right ways to present the turbine efficiency uncertainty bottom line is:

$$U_{ASME} = \pm 0.0081 \text{ at } 0.88 \text{ efficiency}$$

and

$$U_{ASME} = \pm 0.81\% \text{ at } 0.88 \text{ efficiency}$$

Note that the results are clearly percent of a level in both cases. The level is given for reference and the result is clearly associated with it.

Do not get caught with presenting these results wrongly as:

$$U_{ASME} = \pm 0.81\% \text{ at } 88\% \text{ efficiency}$$

The latter is an ambiguous report. Is the 0.81% of 88% or 0.92% in efficiency units, or is it 0.81% in efficiency units at a level of 88% efficiency? The two presentations are clear, the third ambiguous. Lesson number one is, therefore: *Be clearly unambiguous in your presentation.* (Please excuse the oversuperfluity in my redundance.)

Next, it is important to give some information concerning the 0.81% so that a competent technical person (usually not a manager) can better understand the impact of the uncertainty on the test results. Table 8-1 provides the next level of information.

In Table 8-1, the uncertainty model is clearly noted, as are the details of systematic and random uncertainty along with nominal level and even the degrees of freedom. This level of competency in technical affairs (read that as "busy manager") is usually enough. However, the report should also include several more detailed tables of results.

Table 8-1. Turbine Efficiency Systematic and Random Standard Uncertainty Breakdown

	Nominal Level	Systematic Standard Uncertainty (b_R)	Random Standard Uncertainty $(s_{\bar{X}R})$	Uncertainty* $\pm U_{95}$
Efficiency Units	0.88	0.0028	0.0022	0.071
% of Nominal	100	0.32%	0.25%	0.81%
*Note that Uncertainty = $U_{ASME} = \pm 2[(b_R)^2 + (s_{\bar{X}R})^2]$, $v \geq 30$				

Consider Table 8-2, in which the next level of detail is given. Note that throughout, $v = N - 1$, and $N = 1$ (that is, there is only one measurement of each parameter, but there may be many degrees of freedom for the s_X.

Also, for presentation simplicity, all $v \geq 30$. Additional columns may be added for N as appropriate, but they are not included here. In this unit, the emphasis is on the form of the presentation, not the technical detail.

In Table 8-2 the uncertainty model is again noted. Also, the uncertainties cannot be root-sum-squared; they are of different units. The purpose for presenting these data is documentation of the uncertainty magnitudes and measurement levels that went into the uncertainty analysis. A table such as Table 8-2 should be followed by a table such as Table 8-3, in which the uncertainty sources are presented in "results" units; that is, in units of turbine efficiency for this example.

Table 8-2. Turbine Efficiency Elemental Standard Uncertainties in Measurement Units

Uncertainty Source	Units	Nominal Level	Systematic Standard Uncertainty (b)	Random Standard Uncertaint y ($s_{\bar{x}}$)	Number of Points (N)	U_{95}*±
Delta P	psid	4.68	0.00625	0.00245	30	0.013
Airflow	lb/s	147.0	0.53	0.61	1	1.60
Cooling air	lb/s	27.0	0.545	—	1	1.09
Fuel flow	lb/h	15,700.0	22.05	40.5	2	92.1
Burner P	psia	340.0	0.50	0.5	1	1.41
RSS (not needed)	—	—	—	—	—	—

*Note that Uncertainty $= U_{ASME} = \pm 2[(b_R)^2 + (s_{\bar{X}R})^2]$, $v \geq 30$

In Table 8-3, again the uncertainty model is noted. The *RSS* values are shown, as are the final uncertainty, 0.0071, in turbine efficiency units and the degrees of freedom. It is possible to root-sum-square (*RSS*) the systematic and random uncertainties here, because the uncertainty propagation into the same units, result units, has been done.

Table 8-3. Turbine Efficiency Elemental Uncertainties in Result Units

Uncertainty (U_{95})* Source	Systematic Standard Uncertainty (b_R)	Random Standard Uncertainty ($s_{\bar{x}}$)	Degrees of Freedom	Uncertainty ±
Delta P	0.00065	0.00025	≥30	0.0014
Airflow	0.0015	0.00175	≥30	0.0046
Cooling air	0.00195	—	—	0.0039
Fuel flow	0.00055	0.00075	≥30	0.0019
Burner P	0.00105	0.00105	≥30	0.0030
RSS	0.0028	0.0022	≥30	0.0071

*Note that Uncertainty $= U_{ASME} = \pm 2[(b_R)^2 + (s_{\bar{X}R})^2]$, $v \geq 30$

It is important to notice that the data in Table 8-3 show the uncertainty of each measurement in results units. However, the *RSS* of uncertainty is *not done!* The systematic and random uncertainties are root-sum-squared only and then combined to obtain U_{95}. If the uncertainties were root-sum-squared, the result would be 0.0070. This is similar to U_{ASME}, but not the same. It is improper to *RSS* these variable uncertainties. Root-sum-square only the systematic and random uncertainties.

Table 8-3 can be represented as Table 8-4, in which the errors in Table 8-3 are recomputed in terms of percent of nominal turbine efficiency level.

It is important to note that the data in Table 8-4, and in Table 8-3, show the uncertainty of each measurement in results units. However, the *RSS* of the variable uncertainties is *not done!* The systematic and random uncertainties are root-sum-squared only and then combined to obtain U_{95}.

Table 8-4. Turbine Efficiency Elemental Uncertainties in % of Nominal Level Units

Uncertainty (U_{95})* Source	Systematic Standard Uncertainty (b_R)	Random Standard Uncertainty $(s_{\bar{X}})$	Degrees of Freedom	Uncertainty ±
Delta P	0.07385	0.0284	≥30	0.1582
Airflow	0.1705	0.1989	≥30	0.5240
Cooling air	0.2216	—	—	0.4432
Fuel flow	0.0625	0.0853	≥30	0.2115
Burner P	0.1193	0.1193	≥30	0.3374
RSS	0.3190	0.2487	≥30	0.8090
*Note that Uncertainty = $U_{ASME} = \pm 2[(b_R)^2 + (s_{\bar{X}_R})^2]$, $v \geq 30$				

Tables 8-3 and 8-4 are usually the most detailed tables of results a manager would need or like to see. However, there is another way to present the uncertainty results. The data in Table 8-3 can be reformatted to show directly the percent contribution to the total uncertainty by using Equations 8-1 and 8-2, where the percent total uncertainty for each uncertainty term is computed:

$$2b_i \text{ % of total uncertainty} = \{[(2b_i)^2]/[(U_{AMSE})^2]\} \times 100 \qquad (8\text{-}1)$$

$$2s_{\bar{X}, i} \text{ % of total uncertainty} = \{[(2s_{\bar{X}, i})^2]/[(U_{AMSE})^2]\} \times 100 \qquad (8\text{-}2)$$

Equation 8-2 assumes lots of degrees of freedom so that $2s_{\bar{X},i}$ can be used throughout. This is the only case in which these equations can be used properly; they cannot be used for $ts_{\bar{X},i}$.

Using Equations 8-1 and 8-2, Table 8-5, wherein the uncertainty sources are expressed in terms of their percent contribution to the total uncertainty, U_{AMSE}, can be obtained.

Table 8-5. Relative Effect of Uncertainty Sources in % of Total Uncertainty

Error Source	Systematic Part (2b)	Random Part $(2\,s_{\bar{X}})$	Degrees of Freedom	Uncertainty ±
Delta P	3.35	0.50	≥30	—
Airflow	17.85	24.30	≥30	—
Cooling air	30.17	—	—	—
Fuel flow	2.40	4.46	≥30	—
Burner P	8.74	8.75	≥30	—
RSS (not needed)	—	—	—	—

There is no calculation of uncertainty and no calculation of final degrees of freedom in Table 8-5. However, the major uncertainty sources are seen clearly as airflow random uncertainty and cooling air systematic uncertainty. It is here that the major part of any uncertainty improvement program should be concentrated.

8-4.   Summary

Tables 8-1 to 8-5 provide the summary methods for presenting the results of an uncertainty analysis.

Appendix A:
Suggested Reading and Study Materials

APPENDIX A

Suggested Reading and Study Materials

Abernethy, R. B., et al., 1973. *Handbook—Uncertainty in Gas Turbine Measurements*, AEDC-TR-73-5, Arnold AFB, TN: Arnold Engineering Development Center.

ANSI/ASME PTC 19.1-2006, *Instruments and Apparatus, Part 1, Test Uncertainty*.

ANSI/ASME PTC 19.1-1998, *Instruments and Apparatus, Part 1, Test Uncertainty*.

Bowker, A. H., and Lieberman, G. J., 1972. *Engineering Statistics*, 2nd ed., Englewood Cliffs, NJ: Prentice Hall.

Coleman, H. W., Steele, and W. G., Jr., 1998. *Experimentation and Uncertainty Analysis for Engineers*, 2nd Edition, New York: John Wiley & Sons.

Dixon, W. J., and Massey, F. J., 1969. *Introduction to Statistical Analysis*, New York: McGraw-Hill.

Figliola, Richard S. and Beasley, Donald E, 2002, *Theory and Design of Mechanical Measurements*, 3rd Edition, New York, John Wiley & Sons.

Hicks, C. R., 1973. *Fundamental Concepts in the Design of Experiments*, New York: Holt, Rinehart and Winston.

King, J. R., 1971. *Probability Charts for Decision Making*, New York: Industrial Press.

Ku, H. H., 1965. "Notes on the Use of Error Propagation Formulas," NBS Report No. 9011, Washington, DC: National Bureau of Standards.

Ku, H. H., Editor, 1969. *Precision Measurement and Calibration*, NBS Publication 300, Vol. 1, Washington, DC: National Bureau of Standards.

Ku, H. H., Editor, 1969. *Statistical Concepts and Procedures*, NBS Special Publication 300, Vol. 1, Washington, DC: National Bureau of Standards.

Mack, C., 1967. *Essentials of Statistics for Scientists and Technologist*, New York: Plenum Press.

Ross, P. J., 1988. *Taguchi Techniques for Quality Engineering*, New York: McGraw-Hill, NY.

Saravanamutto, H. I. H., Editor, 1990. "Recommended Practices for Measurement of Gas Path Pressures and Temperatures for Performance Assessment of Aircraft Turbine Engines and Components," AGARD Advisory Report No. 245, Neuilly-sur-Seine, France: Advisory Group for Aerospace Research and Development.

Spiegel, M. R., 1992. *Theory and Problems of Statistics*, 2nd ed, New York: Schaum's Outline Series, McGraw-Hill.

Taylor, B. N. and Kuyatt, C. E., *Guidelines for Evaluating and Expressing the Uncertainty of NIST Measurement Results*, NIST Technical Note 1297, 1994 Edition.

Guide to the Expression of Uncertainty in Measurement, 1st Edition, 1993, International Organization for Standardization.

Appendix B: Glossary

APPENDIX B

Glossary

addition uncertainty model—The uncertainty model where the systematic and random uncertainty components are linearly added together. $U_{ADD} = \pm(b_R + t_{95}s_{\bar{X}, R})$

back-to-back tests—Tests run so closely in time that there is no time for instrument drift or calibrations. These tests have zero systematic error and, hence, zero systematic uncertainty.

bad data—Data that disprove "my" theory or thesis.

Benard's formula—The formula used to compute median ranks.

bias error—Old name for systematic error that does not change for the duration of an experiment or test. We now use the term "systematic error."

bias limit—Old name for the estimate of the systematic uncertainty. We now use the term "systematic standard uncertainty."

blunders—Major engineering errors.

calibration errors—Errors resulting from measurement system calibration.

categorize errors and uncertainties—What can be done to ease error analysis bookkeeping.

concomitant functions—Several methods for making the same measurement or calculation that are independent.

confidence—The percentage likelihood that an uncertainty interval, U_{95} or U_{ASME}, about the data average will include the true value.

correlation—The relationship between two data sets. It is not necessarily evidence of cause and effect.

correlation coefficient—The statistic used to calculate correlation.

coverage—A nonstatistic intended to express the likelihood that the true value lies within some interval about the data average. Coverage, and not confidence, is used with U_{ADD} and with U_{RSS}, as these combine a statistic ($s_{\bar{X}, R}$) with a nonstatistic (b_R).

curve fitting—A least-squares method for estimating a curve shape from a data set of X and Y paired values.

curve-fitting cautions—The risks of least-squares curve fitting.

data acquisition errors and uncertainties—A category of error sources and uncertainties related to the method of data acquisition.

data reduction errors and uncertainties—A category of error sources and uncertainties associated with the data reduction process.

degrees of freedom—The amount of room left for error.

delta—The difference between two values.

dithering—The numerical method for evaluating influence coefficients or sensitivity coefficients.

doubter—Personality: "I don't believe in all that uncertainty mumbo-jumbo."

error—(Error) = (measured) – (true). The difference between the measured value and the true value.

Gaussian-normal distribution—The standard expression of the frequency distribution for the most common error data.

good data—Data that prove "my" thesis. Not a proper definition.

good data—Data that can be used for decisions. The proper definition.

group errors or uncertainties—Errors and/or uncertainties grouped into categories of calibration, data acquisition, data reduction, or errors of method.

Grubb's Technique—An outlier identifying technique that identifies few potential outliers.

histogram—Frequency bar chart of a sample of data with the data values as the abscissa and the frequency of occurrence as the ordinate.

indifferent—Personality: "This is not my favorite subject."

influence coefficient—An expression of the influence an error source (and its uncertainty) has on a test result. It is the ratio of the change in the result for an incremental change in an input variable or parameter measured.

interlaboratory—Comparisons made on the same specimen or artifact in order to evaluate the existence and magnitude of systematic error.

least count—The minimum difference between measurements when plotted on probability paper. The smallest increment for a measurement system.

manager—Get the story to him in the first two sentences and make it simple.

matrix analysis—A simple procedure for separating real variation in time and space from random error. The remaining variation is called sampling error and is a random uncertainty.

measurement uncertainty—An interval about the data average that expresses the maximum possible error that may reasonably occur with some confidence. Errors larger than the measurement uncertainty should rarely occur.

median ranks—Using Benard's formula, the percentile rank each data point should occupy if the data sample is normally distributed.

nonsymmetrical systematic standard uncertainty—The systematic standard uncertainty for which there is an uneven likelihood that the true systematic error, β, lies on one side of the data average or the other.

nonsymmetrical standard uncertainty—The uncertainty interval for which there is an uneven likelihood that the true value lies on one side of the data average or the other.

normality—The state of a data set when it is equivalent to a Gaussian-normal distribution.

ostrich—Personality: "I don't want you guys putting fuzz bands on my data."

outliers—"Wild" data; data outside the normally expected data scatter.

outlier rejection—An objective procedure for eliminating outlier data, usually Thompson's τ.

pooling—The combination by squares of several standard deviations, each obtained from a sample from the same population. The pooled standard deviation is a better estimate of the population standard deviation than any one sample.

precision—Old name for random error sources.

precision index—Old name for the standard deviation of the error data divided by the square root of the number of data points to be averaged. $s_{\overline{X}} = s_X / \sqrt{N}$.

presentation—The method for providing the results of an uncertainty analysis in a compelling fashion.

pride—The human factor most to blame for estimating systematic uncertainty too low.

probability plots—A graphical technique for assessing the normality of a data set. Data that plot as a straight line on probability paper are normally distributed.

propagation of error or uncertainty—The analytical technique for evaluating the impact of an error source and its uncertainty on the test result. It employs the use of influence (or sensitivity) coefficients.

random error—An error that may not be predicted from one test reading to the next but whose limits may be estimated with the standard deviation.

random error source—An error source that causes scatter in the test results.

random uncertainty—The standard deviation of error or test data divided by the square root of the number of data points in the average. $s_{\overline{X}} = s_X / \sqrt{N}$

root-sum-square uncertainty model—The uncertainty model where the random and systematic uncertainty components are summed in quadrature. $U_{RSS} = \pm[(b_R)^2 + (t_{95}s_{\overline{X}, R})^2]^{1/2}$

sampling error—A random error source related to the distribution of data in a sample taken from a group of locations such as inside a pipe.

sensitivity—Same as influence coefficient.

standard deviation—Standard deviation of a data set is s_X. $s_X = [\Sigma(X_i - \overline{X})^2/(N-1)]^{1/2}$

standard error of the mean—The standard deviation of a data sample divided by the number of data points averaged.

Student's *t*—The *t* statistic.

symmetrical systematic error—The systematic error type with equal probability of the true systematic error, β, being on either side of the biased average.

systematic error—The error sources that are constant for the duration of the experiment.

systematic standard uncertainty—The estimate of the limits of the systematic error usually estimated with 68% confidence and assumed to have Gaussian-normally distributed error data. It can be either symmetrical or nonsymmetrical, depending on its error source.

Taylor's series error and uncertainty propagation—An error and/or uncertainty propagation method for determining the impact of an error source and/or its uncertainty on the test result.

Thompson's τ technique—An outlier rejection technique that sometimes identifies nonoutlier data as potential outliers.

true value—The desired result of an experimental measurement.

U_{95}—The recommended uncertainty interval with 95% confidence assuming 30^+ degrees of freedom. $U_{95} = \pm t_{95}[(b_R)^2 + (s_{\bar{X},R})^2]^{1/2}$

U_{ASME}—The recommended uncertainty interval with selectable confidence utilizing Student's t for the appropriate degrees of freedom. $U_{ASME} = \pm 2[(b_R)^2 + t_{95}(s_{\bar{X},R})^2]^{1/2}$

U_{ISO}—The selectable confidence uncertainty proposed by the ISO. It is used occasionally in this book, but the uncertainties computed with either U_{95} or U_{ASME} will equal those computed with the ISO model. $U_{ISO} = \pm k[(U_A)^2 + (U_B)^2]^{1/2}$, where k is a constant chosen for the confidence desired, UA is the root-sum-square of the uncertainties ($s_{\bar{X},i}$) for which there are data to calculate a standard deviation, and UB is the root-sumsquare of the uncertainties ($1s_{X,i}$) for which there are no data to calculate a standard deviation. This one-to-one correspondance for U_A and U_B is not necessary for a correct calculation of U_{ISO}.

uncertainty limit—The interval about the data average into which the true value is expected to lie with some stated confidence.

units of error or uncertainty—The expression of the engineering or scientific units for an error or uncertainty analysis. Error and uncertainty units must be the same in order to root-sum-square the systematic and random uncertainties.

weighted degrees of freedom—The degrees of freedom associated with the weighted average.

weighted random uncertainty—The random uncertainty associated with the weighted average.

weighted systematic uncertainty—The systematic uncertainty associated with the weighted average.

weighted uncertainty—The uncertainty limit associated with the weighted average.

weighting—The method of weighting a group of averages of the same test result by the uncertainties of their independent measurement techniques.

Welch-Satterthwaite—The approximation method for determining the number of degrees of freedom in the uncertainty.

Appendix C: Nomenclature

APPENDIX C

Nomenclature

A_t	=	The average of all measurements at time, t, across all locations, l, when performing an analysis of sampling error.
a	=	The first-order constant (slope) in a typical straight line of the formula $Y = aX + b$.
b	=	The systematic standard uncertainty.
b_W	=	The weighted systematic standard uncertainty component of the weighted uncertainty.
b_R	=	$\pm[\Sigma(b_i)^2]^{1/2}$, where b_R and b_i have the same units. b_R is the systematic standard uncertainty of the result.
b_R	=	$\pm\{[\Sigma(\partial F/\partial V_i)^2 \, (b_i)^2]\}^{1/2}$, where b_R is in result units and where the result is obtained with a function $F(V_1, V_2, \ldots V_n)$ with N variables, V_i, each with its own bi in its own units. b_R is the systematic standard uncertainty of the result.
b_R^+	=	The positive component of a nonsymmetrical systematic standard uncertainty.
b_R^-	=	The negative component of a nonsymmetrical systematic standard uncertainty.
b	=	The zero-order constant (intercept) in a typical straight line of the formula $Y = aX + b$.
b_i	=	An estimate of the elemental systematic uncertainty for error/uncertainty source i.
b_i^+	=	The positive component of the nonsymmetrical elemental systematic standard uncertainty for error/uncertainty source i.
b_i^-	=	The negative component of the nonsymmetrical elemental systematic standard uncertainty for error/uncertainty source i.
$d.f.$	=	The degrees of freedom associated with a standard deviation. $d.f.$ is often subscripted.
D_{lt}	=	The measurement at location l and time t, used in a "matrix" analysis of sampling error.
e	=	The base of the natural log where appropriate.
F	=	Degrees Fahrenheit, usually written as °F.
F	=	A function of some variables, usually written $F(V)$.

N = The number of data points or measurements averaged to obtain a measurement of one parameter, such as temperature; N is often subscripted.

P = The symbol for pressure; P is sometimes subscripted.

$P_{0.50}$ = The 50 percentile median rank. It is obtained with Benard's formula: $P_{0.50} = (i - 0.3)/(N + 0.4)$, where i is the ith data point after ordering from lowest to highest value and N is the number of data points in the data set.

psi = Pounds per square inch.

R = A generic result when there is more than one variable or parameter which are combined to yield the result.

r = The correlation coefficient of a data set or sample.

$s_{X,i}$ = The sample standard deviation of error/uncertainty source i.

$s_{\overline{X}}$ = The sample standard deviation of an average for a data set. $s_{\overline{X}, i} = s_X / \sqrt{N}$

$s_{\overline{X}, R}$ = $[\Sigma(s_{\overline{X},i})^2]^{1/2}$, where $s_{\overline{X}, R}$ and $s_{\overline{X}, i}$ have the same units. $s_{\overline{X}, R}$ is the random standard uncertainty of the result.

$s_{\overline{X}, R}$ = $\{[(\partial F/\partial V_i)^2 (s_{\overline{X}, i})^2]\}^{1/2}$, where $s_{\overline{X}, R}$ is in result units and where the result is obtained with a function $F(V_1, V_2, \dots V_N)$ with N variables, V_i, each with its own $s_{\overline{X}, i}$ in its own units. $s_{\overline{X}, R}$ is the random uncertainty of the result.

$s_{X\text{-}CAL}$ = The standard deviation of the calibration data.

SEE = The standard error of estimate. This describes the data scatter around a fitted line. It is analogous to s_X, which describes the data scatter around an average. $SEE = \{[\Sigma(Y_i - Y_{ci})^2/(N - K)]^{1/2}\}$

s_X = The standard deviation of a data set. $s_X = \{[\Sigma(X_i - \overline{X})^2/(N - 1)]^{1/2}\}$

$s_{\overline{X}}$ = The estimate of the standard deviation of the average for a data set of N values. $s_{\overline{X}} = [s_X/(N)^{1/2}]$

$s_{X, pooled}$ = The pooled standard deviation obtained from several estimates of the same population standard deviation.

s_Δ = The standard deviation of the difference between pairs of redundant measurements.

T = The symbol for temperature; T is sometimes subscripted.

TC = The symbol for thermocouple.

t	=	The time of a measurement when performing an analysis of sampling error.
$t_{\alpha,v}$	=	The t-statistic. The uncertainty usually uses 95% confidence, t_{95}, for the appropriate degrees of freedom, v. The usual notation is just t_{95}.
U_R	=	The result uncertainty.
U_W	=	The weighted result uncertainty.
U^+	=	The positive component of a nonsymmetrical uncertainty interval.
U^-	=	The negative component of a nonsymmetrical uncertainty interval.
U_{95}	=	The 95% confidence model uncertainty for any degrees of freedom. $U_{95} = \pm t_{95,v}[(b_R)^2 + (s_{\overline{X},R})^2]1/2$
U_{ADD}	=	The additive model uncertainty. This has _99% coverage. = $U_{ADD} = [(2b)R + t_{95}\, s_{\overline{X},R}]$
U_{ASME}	=	The selectable confidence model uncertainty for 30+ degrees of freedom. $U_{ASME} = \pm 2[(b_R)^2 + (s_{\overline{X},R})^2]^{1/2}$
U_{ISO}	=	The selectable confidence model uncertainty for any degrees of freedom that is used by the International Standards Organization (ISO). $U_{ISO} = \pm k[(U_A)^2 + (U_B)^2]^{1/2}$, where k is a constant chosen for the confidence desired, U_A is the root-sum-square of the standard deviations of the average for error/uncertainty sources for which there are data to calculate a standard deviation ($s_{\overline{X},R}$), and U_B is the root-sum-square of the standard deviations for error/uncertainty sources for which there are no data to calculate a standard deviation ($1s_X$).
U_{RSS}	=	The root-sum-square model uncertainty. This has _95% coverage. $U_{RSS} = \pm[(2b_R)^2 + (t_{95}\, s_{\overline{X},R})^2]^{1/2}$
V	=	A generic variable or parameter being measured; V is often subscripted.
W_i	=	The weight assigned the ith average in computing a weighted average. W_i is also the weight assigned to each random and systematic standard uncertainty when computing the weighted random and systematic uncertainty.
X	=	The generic data point or measurement, a population value; X is often subscripted.
X	=	The abscissa value of a data point (X,Y); X is often subscripted.

$\overline{X}$	=	The average of a data set of X_i data points or measurements.
X_i	=	The ith $\overline{X}$.
Y	=	The ordinate value of a data point (X,Y); Y is often subscripted.
Y_{ci}	=	The Y value of a fitted curve at the point X_i.
α	=	The symbol for the percent coverage for the Student's t distribution, usually taken as 0.95 for 95%.
β	=	The true systematic error for a particular measurement.
γ	=	The ratio of specific heats for a gas.
∂	=	The symbol indicating partial differentiation.
δ	=	$\beta + \varepsilon$ = the true total error for a particular measurement.
δ_{lt}	=	The difference between the measurement at each location, l, and its time, t, average, A_t, when performing an analysis of sampling error.
δ_l	=	The average of all lt at location, l, across time, t, when performing an analysis of sampling error.
Δ	=	The difference between two measurements; Δ is often subscripted.
$\overline{\Delta}$	=	The average of a group of differences between two measurements.
$\overline{\Delta}$	=	The average of the Δ_{lt} values when performing an analysis of sampling error.
Δ_{lt}	=	The values of the residual errors after the linear variation in time and space are averaged out when performing an analysis of sampling error.
ε	=	The true random error of a measurement.
ε_i	=	The difference between a data point Y_i value and a fitted line at the point X_i, i.e., $\varepsilon_i = [Y_i - Y_{ci}] = [Y_i - (aX_i + b)]$.
η_c	=	The symbol for compressor efficiency.
θ_i	=	The influence (or sensitivity) coefficient of the result. It is the effect on the result of error source i (systematic or random).
θ_i	=	$(\partial F / \partial V_i)$.
μ	=	The true average for an entire population.
υ	=	d.f. = degrees of freedom for a standard deviation or SEE; υ is often subscripted. For s_X, $s_{\overline{X}}$, $s_{X,i}$, or $s_{\overline{X},i}$, d.f. = $N - 1$,

where N is the number of data points used to calculate $\overline{X}_i$. For *SEE*, $\upsilon = N - K$ where N is the number of data points in a curve fit and K is the number of constants calculated for the fit.

$\underline{\underline{\upsilon}}$	=	$\underline{\underline{d.f.}}$ = weighted degrees of freedom.
π	=	pi = 3.14159.
ρ	=	The true population correlation coefficient.
σ	=	The standard deviation of an entire population of data (or measurements); sometimes called σ_X.
$\sigma_{\overline{X}}$	=	The standard deviation of the average of a population.
τ	=	Thompson's tau; used to identify outliers with the Thompson's tau technique.

Appendix D: Student's *t* for 95% Confidence

APPENDIX D

Student's *t* for 95% Confidence

v	t_{95}	v	t_{95}
1	12.706	16	2.120
2	4.303	17	2.110
3	3.182	18	2.101
4	2.776	19	2.093
5	2.571	20	2.086
6	2.447	21	2.080
7	2.365	22	2.074
8	2.306	23	2.069
9	2.262	24	2.064
10	2.228	25	2.060
11	2.201	26	2.056
12	2.179	27	2.052
13	2.160	28	2.048
14	2.145	29	2.045
15	2.131	_*	2.000*

For uncertainty analysis, for $v \geq 30$, $t = 2.000$.

Other confidence Student's $t_{\alpha,v}$ tables can be found in most texts on statistics, several of which are listed in Appendix A of this book.

Appendix E: 5% Significant Thompson's τ

APPENDIX E

5% Significant Thompson's τ

Sample Size	τ	Sample Size	τ	Sample Size	τ
3	1.150	16	1.865	29	1.910
4	1.425	17	1.871	30	1.911
5	1.571	18	1.876	31	1.913
6	1.656	19	1.881	32	1.915
7	1.711	20	1.885	33	1.916
8	1.749	21	1.889	34	1.917
9	1.777	22	1.893	35	1.919
10	1.798	23	1.896	36	1.920
11	1.815	24	1.899	37	1.921
12	1.829	25	1.901	38	1.922
13	1.840	26	1.904	39	1.923
14	1.850	27	1.906	40	1.924
15	1.858	28	1.908		

Odds against rejecting a good data point are 20 to 1 or less.

Appendix F:
Areas Under the Normal Curve

APPENDIX F

Areas Under the Normal Curve

Number of σ	Fraction of Area	Number of σ	Fraction of Area
−3.0	0.0014	0.0	0.5000
−2.5	0.0062	0.1	0.5398
−2.3	0.0107	0.2	0.5793
−2.0	0.0228	0.3	0.6179
−1.9	0.0287	0.4	0.6554
−1.8	0.0359	0.5	0.3085
−1.7	0.0446	0.6	0.7257
−1.6	0.0548	0.7	0.7580
_1.5	0.0668	0.8	0.7881
−1.4	0.0808	0.9	0.8159
−1.3	0.0968	1.0	0.8413
−1.2	0.1151	1.1	0.8643
−1.1	0.1357	1.2	0.8849
−1.0	0.1587	1.3	0.9032
−0.9	0.1841	1.4	0.9192
−0.8	0.2119	1.5	0.9332
−0.7	0.2420	1.6	0.9452
−0.6	0.2743	1.7	0.9554
−0.5	0.3085	1.8	0.9641
−0.4	0.3446	1.9	0.9713
−0.3	0.3821	2.0	0.9772
−0.2	0.4207	2.3	0.9893
−0.1	0.4602	2.5	0.9938
		3.0	0.9986

Note that "Number of σ" is the number of standard deviations equivalent to the cumulative area under the normal curve. It is to be used in probability plotting as the linear horizontal axis when probability plotting paper is not available.

Note that "Fraction of Area" is the cumulative area under the curve for the number of σ shown. When probability plotting, the median rank (not multiplied by 100) is equivalent to the "Fraction of Area." Interpolation between the data given is done linearly.

Appendix G: Pressure Instrumentation Selection Example

APPENDIX G

Pressure Instrumentation Selection Example

This is a detailed instrumentation selection example in which pressure instrumentation is selected for a fictional flight test of an aircraft inlet. The pressure instrumentation selected needs to measure dynamic and steady-state pressures in the engine inlet of the aircraft. It is a problem that is closely analogous to measuring pressure profiles and averages in a pipe or duct.

This instrument selection example has six major parts:

1. Define the test setup.

2. Estimate the elemental uncertainties for each pressure transducer type.

3. Root-sum-square the systematic and random uncertainties for each transducer type.

4. Root-sum-square the systematic and random uncertainties from each transducer for each test condition.

5. Compute U_{95}.

6. Repeat analysis to test the effect of proposed improvements.

1. Define the test setup.

For this test, ram recovery (total pressure) will be measured with three calibrated or reference absolute pressure transducers and 39 delta pressure transducers which, when referenced to the three reference transducers, will provide a measurement of the steadystate pressure profile in the engine inlet duct (or in a test pipe). However, for this analysis, it is also necessary to measure inlet distortion. This is done by using the 39 delta pressure transducers to establish the pressure level at each position in the inlet and then measuring pressure fluctuation with high-response transducers at each position.

For this test sequence, two test conditions are of interest. They are summarized as follows. Each typical pressure level or response is noted.

Low altitude, supersonic:

$$P_T = 36 \text{ psia}; \Delta P = 4.4 \text{ psid}; P_{HR} = 0.8 \text{ psi}$$

High altitude, subsonic:

$$P_T = 2 \text{ psia}; \; \Delta P = 0.38 \text{ psid}; \; P_{HR} = 0.3 \text{ psi}$$

where:

P_T = total pressure or ram pressure
ΔP = the typical delta pressure at one of the 39 locations
P_{HR} = high-response pressure

As part of the test setup definition, it is necessary to consider carefully the definition of transducer nonlinearity, end-point fit, hysteresis, and other terms. Figure G-1 illustrates several terms associated with the use of transducer calibration data. These terms are the scatter of the fit, hysteresis, precision (or random uncertainty), nonlinearity, end-point line, and second-order fit.

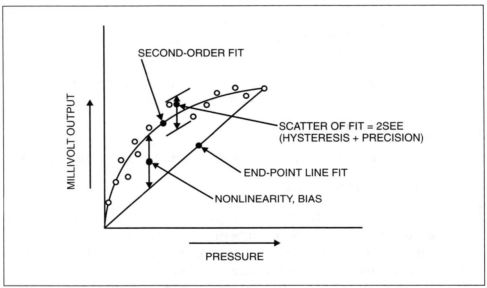

Figure G-1. Typical Transducer Calibration Curve

The scatter of the fit is illustrative of the variation in data about a least-squares secondorder curve fit for a typical transducer. This scatter is due to two physical effects: hysteresis and random uncertainty. The hysteresis is the difference in transducer response experienced when calibration data taken with monotonically increasing pressures are compared with those taken with monotonically decreasing pressures. The precision error effect is seen to cause scatter in the data no matter how the calibration data are taken.

Nonlinearity is a measure of the difference of the true transducer response curve with that of a straight-line fit through the end points of the data, that is, from zero to the maximum data point taken. It is usually the maximum difference possible between the data second-order fit and the end-point straight line.

In this test example, it is necessary to redefine the nonlinearity. Consider Figure G-2, in which a second definition of nonlinearity is presented. The "most common" definition remains presented as in Figure G-1, and the second definition for this test is shown as well. This definition is the difference between the transducer response curve (secondorder fit) and the end-point line, but at the pressure region of interest. In this test, it is necessary to have high-range transducers to ensure they will not be damaged by pressure spikes. Most of the data needed, however, is to be taken at the low end of the transducer performance. Hence, it is necessary to express nonlinearity in the region of interest, not by the usual definition. This is another important lesson in data analysis and uncertainty analysis: *Fit the analysis method to the physics of the test problem; don't try to force the data or methods into a mold.* The physics of the process are all-important.

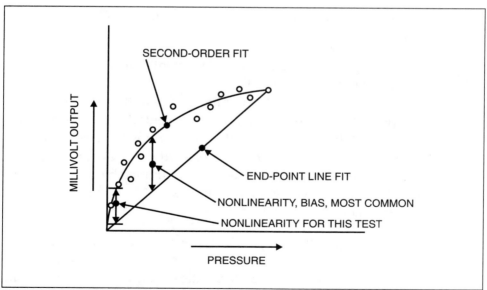

Figure G-2. Low End Nonlinearity Evaluation

It is important to note that in the test setup the ranges of the transducers need to be specified. To start with, these ranges are as follows:

Reference standard transducers (3) = 0–50 psia
Delta pressure transducers (39) = 0–15 psid
High-response transducers (39) = 0–50 psi

2. Estimate the elemental uncertainties for each pressure transducer type.

Now comes the most important and often the most difficult part of any uncertainty analysis. The elemental uncertainties must be estimated for each transducer type. Most of these uncertainties will be obtained from calibration data. Only where necessary will uncertainty estimates be taken from manufacturer's specifications.

(Manufacturer's specifications are often a most misleading source of uncertainty information and should be avoided if at all possible. This author once heard an experimentalist say that when using manufacturer's specifications for uncertainties he takes what the literature says and doubles the uncertainties quoted. He takes what the salesman says and multiplies by ten! Sounds like good advice [tongue in cheek].)

(Note that in all the following calculations, many degrees of freedom are assumed throughout.)

For the 0-50 psia transducer, typical elemental uncertainties are shown in Table G-1. Note that in Table G-1 (and those following) the test conditions are approximated; that is, the nonlinearity is at the approximate test pressure of interest, the zero shift is representative of the transducer in question, and the sensitivity shift represents the transducer response typical of the temperature change expected in the experiment. Note also that the random uncertainties are already divided by the square root of the number of sensors averaged, three in Table G-1 and as appropriate in the tables following.

For the 0-15 psid delta pressure transducer, the elemental uncertainties are as shown in Table G-2. Note that in Table G-2 there are some nonsymmetrical systematic uncertainties. They are not shown to affect the RSS values because they are so small. If they did, the RSS values would also have been nonsymmetrical. Note also that the random uncertainties are divided by the square root of the number of sensors averaged, 39 in this case.

Table G-1. 0-50 psia Reference Standard Transducer Elemental Uncertainties

Uncertainty Source	Systematic, B (psi)	Random $s_{\bar{x}}$ (psi)
A) Calibration	0.005	N/A
B) Nonlinearity (@36 psia)	0 to +0.036	0.0185
C) Hysteresis/Repeatability	N/A	0.0217
D) Zero Shift	0.080	N/A
(0.01% F.S./°F @ +16°F)		
E) Sensitivity Shift		
(0.01% R/°F @ +16°F)		
E1) Low Alt., Supersonic	N/A	0.029
E2) High Alt., Subsonic	N/A	0.016
F) RSS Low Alt., Supersonic	0.088	$0.0405/\sqrt{3} = 0.0235$
G) RSS High Alt., Subsonic	0.088	$0.0285/\sqrt{3} = 0.0165$

Table G-2. 0-15 psid Delta Pressure Transducer Elemental Uncertainties

Uncertainty Source	Systematic, B (psi)	Random $s_{\bar{x}}$ (psi)
A) Calibration	0.0015	N/A
B) Nonlinearity		
B1) Low Alt., Supersonic	0 to +0.0078	0.018
B2) High Alt., Subsonic	0 to −0.009	0.020
C) Hysteresis/Repeatability	N/A	0.0066
D) Zero Shift	0.024	N/A
(0.01% F.S./°F @ +16°F)		
E) Sensitivity Shift		
(0.01% R/°F @ +16°F)		
E1) Low Alt., Supersonic	N/A	0.0035
E2) High Alt., Subsonic	N/A	0.0003
F) RSS Low Alt., Supersonic	0.025	$0.0195/\sqrt{39} = 0.0031$
G) RSS High Alt., Subsonic	0.027	$0.021/\sqrt{39} = 0.0033$

Now note the elemental uncertainties in the 0-50 psi high-response transducers shown in Table G-3. Note here that there is no systematic uncertainty in the elemental uncertainties of a high-response transducer. That is because all the experimenter is interested in is the variation around some mean. Here, the mean is defined as the value of the delta pressure transducer as compared to the reference standard average. Note also that the random uncertainty terms are not divided by the square root of the number of sensors, 39, because there is no averaging. When measuring the dynamic pressure disturbances in a duct, each reading is important and averages have little meaning. Note that to simplify the calculation, throughout this example of many $d.f.$, $B = (2b)$ always, and $S = s$ always.

Table G-3. 0-50 psi High-Response Transducer Elemental Uncertainties

Uncertainty Source	Systematic, B (psi)	Random $s_{\bar{X}}$ (psi)
A) Calibration	N/A	N/A
B) Nonlinearity	N/A	N/A
C) Hysteresis/Repeatability	N/A	0.0225
D) Zero Shift	N/A	N/A
E) Sensitivity Shift (2.2% R @ −65 to +450°F)		
E1) Low Alt., Supersonic	N/A	0.0088
E2) High Alt., Subsonic	N/A	0.0033
F) RSS Low Alt., Supersonic	N/A	0.024
G) RSS High Alt., Subsonic	N/A	0.023

3. Root-sum-square the systematic and random uncertainties for each transducer type.

Note that in Tables G-1, G-2, and G-3, the *RSS* of systematic and random uncertainties has been completed for each transducer type (lines F and G in the tables). Note also that lines F and G represent $s_{\bar{X}}$.

This represents the random uncertainty for each transducer type in Tables G-1, G-2, and G-3. In the uncertainty analysis summary following, it is the value that is root-sumsquared to get the result random uncertainty.

The next step is obtaining the *RSS* value for each systematic and random uncertainty from each transducer type for the test condition in question.

4. Root-sum-square the systematic and random uncertainties from each transducer for each test condition.

5. Calculate U_{95}.

Here the data are summarized in Table G-4 for the total pressure or ram recovery uncertainty. For this uncertainty, it is necessary to recognize that the total pressure (or ram recovery) involves the addition of the average readings from the reference transducers and the average readings from the delta pressure transducers.

In Table G-4, the U_{95} values (lines A4 and B4) represent the uncertainty of total pressure measurement (or ram recovery) at low altitude/supersonic and high altitude/subsonic conditions, respectively.

Table G-5 summarizes the uncertainty analysis for dynamic distortion pressure. Here it is necessary to recognize the test setup results in adding

Table G-4. Total Pressure (or Ram Recovery) Uncertainty

Test Condition	Systematic, B (psi)	Random $s_{\bar{x}}$ (psi)	Table: Line Reference
A) Low Alt., Supersonic			
A1) Reference @ 36 psia	0.088	0.0235	G-1: Line F
A2) Delta P @ 4.4 psid	0.025	0.0031	G-2: Line F
A3) RSS @ 40.4 psia	0.091	0.0235	
A4) $U_{95} = 2[(0.091/2)^2 + (0.047)^2]^{1/2} = 0.102$ psi, or 0.254% R @ 40.4 psia			
B) High Alt., Subsonic			
B1) Reference @ 2 psia	0.088	0.0165	G-1: Line G
B2) Delta P @ 0.38 psid	0.027	0.0033	G-2: Line G
B3) RSS @ 2.38 psia	0.092	0.0169	
B4) $U_{95} = \pm 2[(0.092/2)^2 + (0.0169)^2]^{1/2} = \pm 0.098$ psi, or $\pm 4.1\%$ R @ 2.38 psia			

Table G-5. Dynamic Distortion Pressure Uncertainty

Test Condition	Systematic, B (psi)	Random $s_{\bar{x}}$ (psi)	Table: Line Reference
A) Low Alt., Supersonic			
A1) Delta P @ 4.4 psid	0.025	0.0195	G-2: Line F
A2) High Response @ 0.8 psi	N/A	0.024	G-3: Line F
A3) RSS @ 5.2 psia	0.025	0.031	
A4) $U_{95} = 2[(0.025/2)^2 + (0.031)^2]^{1/2} = 0.067$ psi, or 1.28% R @ 5.2 psia			
B) High Alt., Subsonic			
B1) Delta P @ 0.38 psia	0.027	0.021	G-2: Line G
B2) High Response @ 0.3 psi	N/A	0.023	G-3: Line G
B3) RSS @ 0.68 psia	0.027	0.031	
B4) $U_{95} = \pm 2[(0.027/2)^2 + (0.031)^2]^{1/2} = \pm 0.067$ psi, or $\pm 9.85\%$ R @ 0.68 psia			

the pressure measurements of both the delta pressure transducers and the high-response transducers at each location. No averaging is done.

In Table G-5, the U_{95} values (lines A4 and B4) represent the uncertainty of dynamic pressure measurement at low altitude/supersonic and high altitude/subsonic conditions, respectively.

6. Repeat analysis to test the effect of proposed improvements.

If the uncertainties computed up to now are too large, the uncertainty analysis itself can be used to determine what actions, if taken, would yield improved uncertainty.

First, a shift to a lower-range delta pressure transducer will be evaluated. Then a second-order curve fit will be examined instead of the end-point straight line used in the foregoing analysis. After that, an improved temperature correction for the high-response transducers will be covered.

First, the shift to a lower-range delta pressure transducer is examined. Here the uncertainty of a 0-5 psid delta pressure transducer will be compared to that of a 0-15 psid delta pressure transducer. Consider the uncertainty data in Table G-6, and note that the calibration uncertainty is the same as for the 0-15 psid transducer documented in Table G-2. This is because they have the same calibration source and methods. However, every other uncertainty source is reduced substantially. The resulting RSS uncertainties on lines F and G in Table G-6 objectively illustrate what most experimenters already subjectively believe to be true. That is, lower-range instruments should be used for lower-level measurements.

Table G-6. 0-5 psid Delta Pressure Transducer Elemental Uncertainties

Uncertainty Source	Systematic, B (psi)	Random $s_{\bar{x}}$ (psi)
A) Calibration	0.0015	N/A
B) Nonlinearity		
B1) Low Alt., Supersonic	0 to +0.0026	0.006
B2) High AltT., Subsonic	0 to –0.003	0.065
C) Hysteresis/Repeatability	N/A	0.0022
D) Zero Shift	0.008	N/A
(0.01% F.S./°F @ +16°F)		
E) Sensitivity Shift		
(0.01% R/°F @ +16°F)		
E1) Low Alt., Supersonic	N/A	0.0035
E2) High Alt., Subsonic	N/A	0.0003
F) RSS Low Alt., Supersonic	0.008	$0.0073/\sqrt{39} = 0.0011$
G) RSS High Alt., Subsonic	0.008	$0.0069/\sqrt{39} = 0.0011$

The data in Table G-6 can be summarized in a similar fashion to the data in Table G-2 to yield an uncertainty in both the total (ram) pressure and dynamic pressure readings for the duct. However, the details of that will be left to be reviewed by the student. The resulting impact on the total (ram) and dynamic pressure uncertainties will be summarized later in Table G-9.

Another method for reducing uncertainties in these delta pressure transducers is to use second-order curve fits instead of the end-point straight lines. The impact of this change on the 0-15 psid transducers is shown in Table G-7.

Note that the impact of using the curve fits is to reduce the nonlinearity terms to zero. Note also that, although this is a good idea, it has a negligible impact on the final RSS uncertainty summaries.

Table G-7. 0-15 psid Delta Pressure Transducer Elemental Uncertainties with Second-Order Curve Fits for Each Transducer

Uncertainty Source	Systematic, B (psi)	Random $s_{\bar{x}}$ (psi)
A) Calibration	0.0015	N/A
B) Nonlinearity		
B1) Low Alt., Supersonic	0.0	0.018
B2) High Alt., Subsonic	0.0	0.020
C) Hysteresis/Repeatability	N/A	0.0033
D) Zero Shift	0.024	N/A
(0.01% F.S./°F @ +16°F)		
E) Sensitivity Shift		
(0.01% R/°F @ +16°F)		
E1) Low Alt., Supersonic	N/A	0.0035
E2) High Alt., Subsonic	N/A	0.0003
F) RSS Low Alt., Supersonic	0.024	$0.0195/\sqrt{39} = 0.0031$
G) RSS High Alt., Subsonic	0.024	$0.021/\sqrt{39} = 0.0033$

The next thing that can be tried is to improve on the uncertainty of the high-response transducers by improving their temperature conditioning. This will reduce the impact of the sensitivity shift with temperature. The impact of this change is shown in Table G-8. Here, too, although the change in temperature control is significant, as shown by comparing random uncertainties of the low and high altitudes (lines E1 and E2) in Table G-3 with those in Table G-8, the effect on the uncertainty will be negligible.

Table G-8. 0-50 psi High-Response Transducer Elemental Uncertainties with Improved Temperature Control

Uncertainty Source	Systematic, B (psi)	Random $s_{\bar{x}}$ (psi)
A) Calibration	N/A	N/A
B) Nonlinearity	N/A	N/A
C) Hysteresis/Repeatability	N/A	0.0225
D) Zero Shift	N/A	N/A
E) Sensitivity Shift		
(0.0043% R @ 10°F)		
E1) Low Alt., Supersonic	N/A	0.00018
E2) High Alt., Subsonic	N/A	0.00007
F) RSS Low Alt., Supersonic	N/A	0.0225
G) RSS High Alt., Subsonic	N/A	0.0225

All these changes are summarized in Table G-9 so that the experimenter can decide upon which to do with the money and time available. Some are significant and worth the effort, others are not.

Table G-9. Illustration of Uncertainty Improvements with Changes in Data Analysis, Selection, and Us

	Uncertainty			
	Low Alt., Supersonic		High Alt., Subsonic	
Measured Pressure	**psi**	**(% R)**	**psi**	**(% R)**
Total pressure (ram)				
Baseline	±0.101	(±0.25%)	±0.098	(±4.1%)
5 psid range ΔP	±0.100	(±0.25%)	±0.098	(±4.1%)
Second-order fits (15 psid ΔP)	±0.102	(±0.25%)	±0.099	(±4.2%)
Dynamic distortion				
Baseline	±0.067	(±1.28%)	±0.067	(±16.3%)
5 psid range ΔP	±0.049	(±0.94%)	±0.049	(±12.0%)
Second-order fits (15 psid ΔP)	±0.067	(±1.28%)	±0.067	(±16.3%)
Better HR temp. control	±0.065	(±1.25%)	±0.065	(±16.3%)

As can be seen in Table G-9, the only action that has any real effect is to go to second-order curve fits, and that improved only the dynamic distortion pressure measurements. It should be done, however, if the need for the accuracy improvement—that is, reduced uncertainty—is there.

The above gory details show that uncertainty analysis can be used to evaluate the potential impact of measurement system changes on the uncertainty before proceeding with the expensive task of implementing them.

Appendix H:
Alternate Measurement Uncertainty Analysis Approach

APPENDIX H

Alternate Measurement Uncertainty Analysis Approach

It is important to recognize that the uncertainty analysis methods proposed by the Bureau International de Metrologie Legale (BIPM) and the International Standards Organization (ISO), Technical Advisory Group (TAG) 4, Working Group (WG) 3, yield the same result uncertainty as the methods presented in this book.

This text and the BIPM/ISO/TAG4/WG3 guidelines have many similarities. Their primary differences are in the designation and combination of error and uncertainty estimates. The BIPM/ISO/TAG4/WG3 guidelines divide error sources into two groups: Type A error sources are those error sources for which there are data to calculate a standard deviation; Type B error sources are all the other error sources. This text and the more universally accepted approach of ASME, ISA, several ISO documents, and others, group errors as either bias (systematic) or precision (random). In addition, this text utilizes subscripts, A or B, to denote the ISO uncertainty or error source type. In this way, this ILM combines the best of both worlds.

Most users of measurement uncertainty analysis data don't care much about the origin of the error classification—the BIPM/ISO/TAG4/WG3 guidelines approach. They care much more about the effects of the errors; that is, do they shift the data from the true value or do they cause scatter in the test results? The approach taken by this book handles both needs well.

While the BIPM/ISO/TAG4/WG3 guidelines offer a very simple way to group error sources, the groups have little or no meaning to the individual who faces a test or experiment where the errors are too large. The BIPM/ISO/TAG4/WG3 guidelines do not provide any help with discerning where errors should be reduced because they provide no view into the effect of error sources.

The latter is a major strength of the methods in this book. If an experiment or test has a measurement uncertainty that is too large, the measurement uncertainty analysis identifies the major error source drivers as either systematic or random. The approaches to reducing the effects of these two error types are very different. If systematic uncertainty is the problem, the experimenter needs better instrumentation or better calibrations. If random uncertainty is the problem, securing more data will likely help.

Neither of these diagnostic approaches is available with the BIPM/ISO/ TAG4/WG3 guidelines.

It should be mentioned also that the BIPM/ISO/TAG4/WG3 guidelines treat every error source as though it were one standard deviation of the average and then rootsum-square all the estimates of their limits together for the total uncertainty. There is no recommended confidence or coverage for the total uncertainty interval, that choice being left to the user.

The methods of this book agree with both the U.S. National Standard (ASME PTC 19.1) and the ISO Guide. The 95% confidence model is preferred.

Appendix I: Solutions to All Exercises

APPENDIX I

Solutions to All Exercises

Unit 2

2-1. Student's t

  a. $s_{\overline{X}} = s_X/(N)^{1/2}$

  b. Random standard uncertainty is $s_{\overline{X}} = \dfrac{s_X}{\sqrt{N}}$

  c. 95% of the time, μ will be contained in the interval
      $(\overline{X} - t_{95}s_X/(N)^{1/2}) \le \mu \le (\overline{X} + t_{95}s_X/(N)^{1/2})$

  d. 1) Random uncertainty component $= t_{95}s_X/(N)^{1/2} =$
      $(2.20 \times 1.7)/(12)^{1/2} = 1.08$
      2) See (c).

2-2. Pooling

  a. 301 (most data)

  b. $2.131 \times 325/(16)^{1/2} = 173$ $3.182 \times 280/(4)^{1/2} = 445$
      $2.179 \times 297/(13)^{1/2} = 179$ $2.365 \times 291/(8)^{1/2} = 243$
      $2.074 \times 301/(23)^{1/2} = 130$
      (Remember, $N = \upsilon + 1$)

  c. 304; 59

2-3. Dependent Calibration

  a. Separate calibration (in the same calibration facility) will
      reduce the calibration random error by averaging but still cost
      more. An uncertainty analysis and a cost analysis are needed to
      determine the cost effectiveness. With the high cost of gasoline,
      extraordinary measures are being taken to improve the accu-
      racy of measuring flow.

  b. A student once said, "It depends upon whether or not my
      cousin owns the independent calibration laboratory." How-
      ever, the right answer is: Independent calibrations (different
      labs) would reduce both calibration systematic and random
      uncertainty. An uncertainty analysis and a cost analysis are
      needed.

  c. This will produce "linkage" or dependency in the two flows.

d. In the general case, there are many different input and output flows. A weighted average based on the random uncertainties (variances) of the flow rates is often employed. Least-squares regression is used to establish the weighting factors. Some writers have suggested weighting by uncertainties rather than by random uncertainties only. This is described in Unit 6.

Unit 3

3-1.   Systematic Uncertainty

a. $b = (\sum_i b_i^2)^{1/2} = 10.68$

b. 19.47

c. Large uncertainties dominate the calculation of the uncertainty of the result.

3-2.   Random Uncertainty

a. $s_{\overline{X}} = (1^2 + 2^2)^{1/2} = 2.24\%$

b. $v = \dfrac{(1^2 + 2^2)^2}{(1^4/10) + (2^4/20)} = \dfrac{25}{(2 + 16)/20} = 27.8$

(Truncate for conservative answer)
$v = 27$

c. $t_{95,\ 27} = 2.052$ (Note: the subscript 27 is for the degrees of freedom)

$t_{95}s_{\overline{X},\ R} = 2.052 \times 2.24\% = 4.60\%$

If the systematic error (or uncertainty) is zero or negligible, about 95% of the time the true value should lie within the interval of the average plus or minus $t_{95}s_{\overline{X},\ R}$.

3-3.   Combining Degrees of Freedom

Just sum over i to obtain $s_{\overline{X},\ R} = [\Sigma(s_{\overline{X},\ i})^2]^{1/2}$.

3-4.   Truncating Degrees of Freedom

Truncating yields a lesser degrees of freedom with an associated larger t_{95}. This results in a slightly larger random error component of the uncertainty and slightly larger uncertainty. Thus, this overestimating of uncertainty is a safer (more conservative) path for decisions.

3-5.   Assuming all degress of freedom are 30+ we have:

Calculate $q = [((26 + 1.5) + (26 - 3.5))/2] - 26 = -1$ (the nonsymmetrical systematic offset).

Calculate $b = [(27.5 - 22.5)/2] = 2.5$ (the symmetrical systematic standard uncertainty)

$U_{95} = \pm 2[(2.5)^2 + (4)^2]^{1/2} = \pm 9.4$.

Calculate the 95% confidence symmetrical uncertainty interval:
Interval (95%) = $(26 + (-2)) \pm 9.4$
$U^+ = 9.4 - (-2) = 11.4$
$U^- = 9.4 + (-2) = 7.4$

Calculate the upper and lower nonsymmetrical uncertainties:
$I_{upper} = 26 + [9.4 + (-2)] = 33.4$
$I_{lower} = 26 - [9.4 - (-2)] = 14.4$

The uncertainty interval is [14.6 to 33.4] or $\overline{X}$ +7.4, −11.4.

3-6.   Scale and Truth

a.   Truth in this case is the reading on the doctor's scale.

b.   Consideration

c.   and d. Consideration

Unit 4

4-1.   Terminology

a. ε is the real, true random error in any one measurement. $2s_X/\sqrt{N}$ is the random uncertainty for large degrees of freedom, >30.

The distribution of error it represents is caused by the the true error, ε, in each measurement.

b. β is the true systematic error for any one measurement or group of measurements in a process.

B is the systematic uncertainty; it is the estimate of the limits possible for the true systematic error, β.

c. δ is the true total error in one measurement. It is $(\varepsilon + \beta)$.

U is the measurement uncertainty, usually noted as U_{95}. It is the estimate of the limits possible for the total error with some confidence.

4-2.   Pooling Review

a. The $s_{X,i}$ values represent three estimates of the population standard deviation,

b. 50.0; it is the $s_{X,i}$ with the most data.

c. 43.0

d. 34

4-3.   Delta

a. 1.72 and 1.51, respectively.

b. No, they also include process variations.

c. –0.9, –1.0 and –0.6

d. 0.21

e. $0.21 / \sqrt{2} = 0.15$

f. Two measurement systems can be used to evaluate the standard deviation of either if they are measuring the same process. Using only one would yield a standard deviation larger than that caused by the measurement.

4-4.   Interfacility Results

a. The best estimate available for the orifice flow.

b. $\pm s_X = \pm 0.2$ lb/s

c. (b) represents the scatter between laboratories. The closeness of the 12.0 lb/s to the true value is obtained by first dividing (b) above by $\sqrt{7}$ to obtain the random standard uncertainty. The one multiplies by Student's t of 2.447, yielding 0.18 lb/s. That is, 12 ± 0.18 lb/s contains the true value 95% of the time considering only random components of error or uncertainty.

d. Each laboratory may interpret (b) above as an estimate of the systematic standard uncertainty between laboratories or the systematic uncertainty to assign each one.

4-5.   Systematic Standard Uncertainty Impact

a. 66

b. No. It assumes all the elemental error components work to their limits in the same direction at the same time. That is not realistic.

c. 60.1

d. No. Same as (b) above for the largest, most influential elemental errors.

e. 37.6

f. Yes. All elemental error components have the chance to cancel, and this calculation considers that.

g. (e)

4-6.   Back-to-Back

a. $U_{CAL} = 2[(b_{CAL})^2 - (s_{\bar{X}-CAL})^2]^{1/2}$ (fossilized)
    $= 2[((0.5)1/2^2 + ((0.5)\,1/2)^2]^{1/2} = 1.58\%$
    (all systematic)

b. $b_{CAL} = 0.707\%$, $s_{\bar{X}-CAL} = 0.707\%$

c. No calibration error!

d. Correct for documented drift in time. This would increase the above errors by the error in designating the drift in time.

4-7.   Observable Systematic Error

a. Unknown. The $\pm 0.5\%$ is only the random component of the uncertainty. The systematic standard uncertainty is unknown, so the uncertainty is unknown.

b. No. A consistent shift of all the data cannot be observed; only the scatter in the data can be observed.

c. It could be large but can't be calculated. The uncertainty is unknown. If the systematic standard uncertainty is large, there is great risk attendant to any decision based on these test results.

4-8.   Compressor Uncertainty

a. There is a great risk in misinterpreting the test results. The test should be postponed until the uncertainty can be improved. Review the uncertainty analysis to determine:

1) Systematic and/or random problems
2) Which measurand is involved, or are there several?

From these answers, analyze the alternatives to select the most cost-effective improvement.

b. A simple view would be $\pm 2s_{\bar{X}} = \pm 1.2\%$, i.e., close enough!

c. (1) Take more data points and average, (2) use more redundant instrumentation for the most imprecise measurand, or (3) obtain a more precise measuring probe, transducer, or instrument.

d. 1) Better calibration method or standard
2) More accurate instrumentation
3) Independent calibration of redundant instruments
4) Concomitant variable

4-9.   Historic Judgment Answers:

a-e. Tests should be designed to "average out" these sources if a cost-uncertainty trade study shows this to be effective. If the cost trade is not effective, systematic uncertainties must be added to account for the risk that a single test will produce optimistic results. A fossilized systematic uncertainty, $2[(b/2)^2 + (s_{\bar{X}})^2]^{1/2}$, would be used if known, otherwise a systematic uncertainty comes from subjective judgment.

4-10.   *"Why measure that way" Answers:*

a.

	Idle	Max
CO	F	P
HC	F	P
NO_X	P	F

Everything measured over the limit fails as does every point that is over when the uncertainty is added to the measurement. (Compliance has not been demonstrated.)

b.

	Idle	Max
CO	P	P
HC	P	P
NO_X	P	F

Everything under the limit passes as does everything when the measurement minus the uncertainty is under. (Failure has not been proved.)

c. Yes. CO idle and HC idle. It depends upon whom the burden of proof lies as to how the uncertainty is applied.

d. Better HC accuracy may remove the regulator's doubts. Break up of systematic and random uncertainties may allow repeat tests to lower uncertainty and pass the car.

e. Actual levels are likely not different from zero. (The uncertainty exceeds the measurement).

Unit 5

5-1. Purpose of Uncertainty Propagation

Uncertainty propagation is needed to combine the effects of errors (the uncertainties) from several sources and to evaluate their effects on the test or experimental result. It is not possible to combine uncertainties in temperature and pressure, for example, into their impact on flow until the temperature and pressure uncertainties are converted into the result units of flow. This is what uncertainty propagation does.

5-2. Orifice

a. $s_{P_{up}} = 2.8835$; $s_{P_{dn}} = 2.9925$

b. 1) $s_{\Delta Pi} = [(1)^2(s_{P_{up}})^2 + (-1)^2(s_{P_{dn}})^2]^{1/2}$

2) $s_{\Delta Pi} = [(2.8835)^2 + (2.9925)^2]^{1/2} = 4.1557$

3) $s_{\Delta Pd} = [(1)^2(s_{P_{up}})^2 + (-1)2(s_{P_{dn}})^2 + 2(1)(-1)r(s_{P_{up}})(s_{P_{dn}})]^{1/2}$

4) Use formula in (3). Remember, $r = 0.9983$.

$s_{\Delta Pd} = [(2.8835)^2+(2.9925)^2 - 2(0.9983)(2.8835)(2.9925)]^{1/2} = 0.203$

5) Just list the differences between each upstream and its corresponding downstream pressure.

6) $s'_{\Delta P} = 0.201$

7) The correct $s_{\Delta P}$ is answer 6). There the actual delta pressure data have been used to calculate the standard deviation of those deltas. However, answer 4) is also correct as the uncertainty propagation has properly considered the dependency, or lack of independence, of the delta pressure measurements.

Answers 6) and 4) agree as proof of the Taylor's Series uncertainty propagation method. Answer 2) is dead wrong. Included in its calculation of the standard deviation of the delta pressures is the standard deviation of some effect that is shifting both the upstream and downstream pressure measurements. Those measurements are correlated and can't be assumed to be independent as is done in the usual Taylor's Series uncertainty propagation. The lesson? Watch out for non-independent uncertainty sources. You can get killed!

5-3.   Word Problem

When airflow is contained explicitly in both gross thrust and drag equations, it is not correct to first propagate the uncertainties in airflow into uncertainty in gross thrust and then follow (by propagation) the uncertainties in airflow into drag, finishing by propagating the uncertainties in gross thrust and drag into net thrust. In that case, the uncertainties in airflow are linked to the uncertainty propagation results obtained for both gross thrust and drag. Airflow uncertainty is implicit in the propagation of gross thrust and drag uncertainties into the final result of net thrust. This will not properly account for the correlation that then results. What is necessary is to propagate the uncertainty in airflow, gross thrust, and drag in one expression through to net thrust. In that case, the linked uncertainty of airflow will be explicitly present in the net thrust equation and properly handled by the Taylor's Series uncertainty propagation.

5-4.   Degrees of Freedom

The Welch-Satterthwaite equation for combining degrees of freedom is given as Equation 3-4 for this simple, all-the-same-error category, exercise.

$$\upsilon = [\Sigma(s_j)^2]^2 / \{\Sigma[(s_j)^4 / \upsilon_j]\}$$

a. For the first random root-sum-square uncertainty in Table 5-2, Equation 3-4 becomes:

$$\upsilon_{0.57} = \frac{[(0.20)^2 + (0.20)^2 + (0.02)^2 + (0.50)^2]^2}{[(0.20)^4/45 + (0.20)^4/8 + (0.02)^4/3 + (0.50)^4/20]}$$

$$= [0.109164]/[0.003361] = 32.48$$

$$\cong 32 \text{ (always truncate)}$$

For the other RSS random uncertainty in Table 5-2, Equation 3-4 becomes:

$$\upsilon_{0.67} = \frac{[(0.57)^2 + (0.30)^2 + (0.20)^2]^2}{[(0.57)^4/32 + (0.30)^4/16 + (0.20)^4/1]}$$

$= [0.206934]/[0.005405] = 38.29$

$\cong 38$ (always truncate)

b. Undefined. The units of the random uncertainties are not the same (pressure and temperature) so they can't be root-sum-squared until after uncertainty propagation. (This was a trick question.)

5-5. Parallel Flowmeters Uncertainty Propagation

Assuming equal flow in each of the meters, the equation for the total flow, F, is:

$$F = A + B + C$$

where A, B, and C are the flows in each of the three meters. Using Equation 5-16, the following uncertainty propagation equation is obtained for systematic standard uncertainty. Simply substituting $s_{\bar{X}}$ for b in this equation will yield the uncertainty propagation equation for random standard uncertainty.

$$\begin{aligned} b_F = [(\partial F/\partial A)^2(bA)^2 &+ (\partial F/\partial B)^2(bB)^2 + (\partial F/\partial C)^2(bC)^2 \\ &+ 2(\partial F/\partial A)(\partial F/\partial B)(b_A)(b_B) \\ &+ 2(\partial F/\partial C)(\partial F/\partial B)(b_C)(b_B) \\ &+ 2(\partial F/\partial A)(\partial F/\partial C)(b_A)(b_C)]^{1/2} \end{aligned}$$

This is the uncertainty propagation equation for absolute units. It will also work for relative units, % terms, but the sensitivities or influence coefficients will be more complicated.

5-6. Decision Time

a. For all independent uncertainties (errors), all the correlation coefficients in the uncertainty propagation equation, r, are 0.0 and the three cross-product terms go to zero. The numerical values of the partial derivative terms, the influence coefficients, or sensitivities are all 1.0 by inspection. If all the systematic and random standard uncertainties are 1.0 gal/min, the test or

experimental systematic and random standard uncertainties
are calculated as follows:

$$b = [(1.0)^2(1.0)^2 + (1.0)^2(1.0)^2 + (1.0)^2(1.0)^2]^{1/2}$$
$$= [3.0]^{1/2} = 1.7 \text{ gal/min}$$

$$s_{\overline{X}} = [(1.0)^2(1.0)^2 + (1.0)^2(1.0)^2 + (1.0)^2(1.0)^2]^{1/2}$$
$$= [3.0]^{1/2} = 1.7 \text{ gal/min}$$

Equation 5-21 is the uncertainty propagation equation for three
meters in series. By inspection and comparison to the above, it
is seen that for the series case the systematic standard uncertainty is one-third the above, as is the random standard uncertainty. Therefore, for independent errors (or uncertainties) the
series network is more accurate.

b. For the case where the systematic errors (and uncertainties) are
completely dependent, the correlation coefficients, r, in the
above general uncertainty propagation equation are 1.0. The
following is then obtained for systematic standard uncertainty:

$$b = [(1.0)^2(1.0)^2 + (1.0)^2(1.0)^2 + (1.0)^2(1.0)^2$$
$$+ 2(1.0)(1.0)(1.0)(1.0)(1.0)$$
$$+ 2(1.0)(1.0)(1.0)(1.0)(1.0)$$
$$+ 2(1.0)(1.0)(1.0)(1.0)(1.0)^{1/2}$$
$$= 3.0$$

The same answer is obtained for random uncertainty as
obtained above, that is, $s_{\overline{X}} = 1.7$, since random errors from one
time to another can never be dependent.

For the series case, the systematic standard uncertainty is, by
inspection,

$$b = 3.0/3.0 = 1.0.$$

The series case is still more accurate, and correlated systematic
error (or uncertainty) has increased the systematic standard
uncertainty. This does not always happen, but the Taylor's
Series approach will give the correct answer for every circumstance.

5-7.   First Impression Answers:

a. By inspection: Random $(s_{\overline{X}})T_2$ is largest; Systematic $(b)T_1$ is
smallest.

b. $\dfrac{\partial Q}{\partial T_1} = 2500; \dfrac{\partial Q}{\partial T_2} = 625$

   i. $b = \left[\left(\dfrac{\partial Q}{\partial T_1}\right)^2 (b_{T_1})^2 + \left(\dfrac{\partial Q}{\partial T_2}\right)^2 (b_{T_2})^2\right]^{1/2}$

          $= (2500^2 * 3^2 + 625^2 * 5^2)^{1/2}$

          $= (56{,}250{,}000 + 9{,}765{,}625)^{1/2} = 8125$

   ii. Random $= \left[\left(\dfrac{\partial Q}{\partial T_1}\right)^2 (s_{T_1})^2 + \left(\dfrac{\partial Q}{\partial T_2}\right)^2 (s_{T_2})^2\right]^{1/2}$

          $= (2500^2 * 3^2 + 625^2 * 4.5^2)^{1/2}$

          $= (56{,}250{,}000 + 7{,}910{,}156.3)^{1/2} = 16020$

   iii. $U_{95} = 2[(b)^2 + (s_{\bar{X}})^2]^{1/2} = 17105 = \pm 54.7\%$

c.

56,250,000	56,250,000
9,765,625	7,910,156.5

   i. Two are the largest, T_1 systematic and random (b and $s_{\bar{X}}$), smallest is T_2 systematic (b).

   ii. ————————

   iii. Influence coefficients, or sensitivity factors, really matter.

5.8.   Relative Effect of Uncertainty Sources: Derivation Method Answers:

a. Equations

   i. Note each systematic term is the following fraction of (b_{RSS}):

$$\dfrac{(b_i)^2}{(b_{RSS})^2}$$

   ii. That term is the following fraction of Total Uncertainty (TU):

$$\dfrac{(b_{RSS})^2}{(TU)^2}$$

iii. Therefore % of Total Uncertainty (TU) for each uncertainty source, i, is:

$$Syst.\%TU = \left(\frac{(b_i)}{(b_{RSS})^2}\right)\left(\frac{(b_{RSS})^2}{(TU)^2}\right)(100) = \left(\frac{(b_i)^2}{(TU)^2}\right)(100)$$

iv. Similarly:

$$Rand.\%TU = \left(\frac{(2s_i)^2}{(TU)^2}\right)(100)$$

b. Proof that sum is unity (100%) assuming 30+ degrees of freedom

For $(2b_1 = B_1)$ and $(2S_1)$
$(2b_2 = B_2)$ $\quad$ $(2S_2)$
$(2b_3 = B_3)$ $\quad$ $(2S_3)$

$$B_1\%TU = \left(\frac{(B_1)^2}{(TU)^2}\right)(100) = \frac{(B_1)^2(100)}{((B_1)^2 + (B_2)^2 + (B_3)^2) + ((2S_1) + (2S_2) + (2S_3))}$$

For the sum:

$$B_1\%TU + B_2\%TU + B_3\%TU + 2S_1\%TU + 2S_2\%TU + 2S_3\%TU$$

$$Sum = \frac{((B_1)^2 + (B_2)^2 + (B_3)^2) + ((2S_1) + (2S_2) + (2S_3))(100)}{((B_1)^2 + (B_2)^2 + (B_3)^2) + ((2S_1) + (2S_2) + (2S_3))} = 100 \quad \text{q.e.d.}$$

Unit 6

6-1.   Why Weight

a. A weighted result will be more accurate (lower uncertainty) than any of the methods being weighted.

b. All weighting formulas must have the same units for all the uncertainty terms. That is, all systematic uncertainty terms and all random uncertainty terms must have the same units, usually those of the test result.

6-2.   Weighting by Uncertainty

a. $\bar{X}_G = \dfrac{\sum N_i \bar{X}_i}{\sum N_i} = 97.7$

b. $W_2 = \dfrac{(U_1 U_3)^2}{W_T}; W_T = [(U_2 U_3)^2 + (U_1 U_3)^2 + (U_1 U_2)^2]$

$W_3 = \dfrac{(U_1 U_2)^2}{W_T}$

c. $W_1 = \dfrac{15804.997}{15804.997 + 5799.997 + 4360.004} = 0.60870$

$W_2 = 0.22338$

$W_3 = 0.16792$

Check: $W_1 + W_2 + W_3 = 1$ if correct.

d. $\bar{X}_W = 95.56; U_{95} = 2[(b)^2 + (s_{\bar{X}})^2]^{1/2}$

$b_{R,W} = [(0.60870 \times (3))^2 + (0.22338 \times (5))^2 + (0.16792 \times (4))^2]^{1/2}$
$= 4.487$

$s_{\bar{X},R,W} = [(0.60870 \times 1)^2 2 + (0.22338 \times 1.5)^2$

$+ (0.16792 \times 4.5)^2 2]^{1/2} = 1.027$

$U_{95,W} = \pm 2[(4.487)^2 + (1.027)^2]^{1/2} = \pm 4.93$

e. Weighting produces a more accurate average, with smaller uncertainty.

Unit 7

7-1.   Outliers: Thompson's tau (τ)

a. 1. $\bar{X} = 14.158; s = 120.7; N = 19$
2. 334 is a suspect outlier.
3. $\delta = |334 - 14.158| = 319.842$
4. $\tau = 1.881$
5. $\tau s = 1.881 \times 120.7 = 227.0$
6. $\delta > \tau s$ [319.842 > 227.0]. Therefore, 334 is an outlier.

b. 1. $\bar{X} = -3.611$; $s = 95.3$; $N = 18$
    2. -220 is a suspect outlier.
    3. $\delta = |-220 - (-3.611)| = 216.389$
    4. $\tau = 1.876$
    5. $\tau s = 1.876 \times 95.3 = 178.8$
    6. $\delta > \tau s = [216.4 > 178.8]$. Therefore, -220 is an outlier.

c. 1. $\bar{X} = 9.118$; $s = 80.9$; $N = 17$
    2. 166 is a suspect outlier.
    3. $\delta = |166 - 9.118| = 156.882$
    4. $\tau = 1.871$
    5. $\tau s = 1.871 \times 80.9 = 151.4$
    6. $\delta > \tau s [156.882 > 151.4]$. Therefore, 166 is an outlier.

d. 1. $\bar{X} = -0.688$; $s = 72.4$; $N = 16$
    2. 129 is a suspect outlier.
    3. $\delta = |129 - (-0.688)| = 129.688$
    4. $\tau = 1.865$
    5. $\tau s = 1.865 \times 72.4 = 135.026$
    6. $\delta > \tau s [129.688 < 135.026]$. Therefore, 129 is *not* an outlier!

7-2. Curve Fitting

a. For CO_2—A, order 3
    For CO_2—B, order 5

b. No, there may not be enough data for these third- or fifth-order fits.

c. All CO_2—A ranges: $0.028 \approx 0.044$ (0–5%) and
    $0.020 \approx 0.022$ (0–2%); close enough.

Two CO_2—B ranges: $0.101 \approx 0.115$ (0–18%) and
$0.035 \approx 0.044$ (0–5%); close enough.

The 2SEE scatter in the above approximates the standard calibration gas error.

7-3. Least-Squares Curve Fit

$$\Delta = y - Ax^2 - Bx$$
$$\Delta^2 = y^2 - 2Ayx^2 - 2Byx + A^2x^4 + 2ABx^3 + B^2x^2$$
$$\Sigma\Delta^2 = \Sigma y^2 - 2A\,\Sigma yx^2 - 2B\,\Sigma yx + A^2\,\Sigma x^4 + 2AB\,\Sigma x^3 + B^2\,\Sigma x^2$$

Minimize $\Sigma\Delta^2$ with respect to A and B:

$$\frac{\partial(\Sigma\Delta^2)}{\partial A} = -2\Sigma yx^2 + 2A\Sigma x^4 + 2B\Sigma x^3 = 0$$

$$\frac{\partial(\Sigma\Delta^2)}{\partial B} = -2\Sigma yx + 2A\Sigma x^3 + 2B\Sigma x^2 = 0$$

Solve for A and B:

$$A = \frac{\Sigma x^2 \Sigma yx^2 - \Sigma yx \Sigma x^3}{\Sigma x^2 \Sigma x^4 - (\Sigma x^3)^2}$$

$$B = \frac{\Sigma yx - A\Sigma x^3}{\Sigma x^2}$$

7-4. Correlation

   a. At 95% $t \cong 18.7$. Yes, r_A is significant

   b. At 95% $t \cong 4.1$. Yes, r_B is significant

   c. At 95% $t \cong 2.6$. Yes, r_C is significant

   d. This is simple X versus Y plots.

   e. Two outliers in data set B; eliminate them.

   f. $r_B = -0.009$; $t \cong 0.03$. No, the new r_B is not significant

   g. Even high correlation coefficients (close to ±1.0) are not always significant.

7-5. Probability

   a. Just line up the ordered data points with the ordered median ranks.

   b. Plots

   c.

Data Set	Line	Conclusion	$\bar{X}$	s_X	Outliers?
A	Straight	Normal Data	~ -0.65	~ 1.18	No
B	Straight	Normal Data	~ 0.17	~ 0.61	One
C	Two	Bimodal	—	—	No

d.

Data Set	Calculated	
	$\bar{X}$	s
A	−0.647	1.123
B	0.064	0.797
C	1.792	2.204

e. Rework data set *B* after outlier is removed. Determine source of bimodality in data set *C*.

Index

PRIME EQUATIONS FOR UNCERTAINTY ANALYSIS

Random standard uncertainty equations:

standard deviation

$$s_X = \sqrt{\frac{\sum\limits_{i=1}^{N} (X_i - \overline{X})^2}{N-1}}$$

random standard uncertainty (one ave.)

$$s_{\overline{X}} = \frac{s_X}{\sqrt{N}}$$

interval that contains the true value, μ, in the absence of systematic errors

$$\overline{X} - t s_{\overline{X}} < \mu < \overline{X} + t s_{\overline{X}}$$

random standard uncertainty (multiple averages)

$$s_{\overline{X}} = \left[\frac{\sum\limits_{i=1}^{M} (\overline{X}_i - \overline{\overline{X}})}{M-1}\right]^{1/2}$$

combined effects of several random standard uncertainties

$$s_{\overline{X}} = \left[\sum\limits_{i=1}^{N} (s_{\overline{X},i})^2\right]^{1/2}$$

degrees of freedom, random standard uncertainty only

$$d.f. = v = \frac{\left[\sum\limits_{i=1}^{N} (s_{\overline{X},i})^2\right]^2}{\left[\sum\limits_{i=1}^{N} \frac{(s_{\overline{X},i})^4}{v_i}\right]}$$

Systematic standard uncertainty equations:

For **systematic standard uncertainty**, assume: b represents the 68% confidence interval for normally distributed error data, and, b has infinite (∞) degrees of freedom.

combined effects of several systematic standard uncertainties:

$$b_R = \left[\sum\limits_{i=1}^{N} (b_i)^2\right]^{1/2}$$

Recommended uncertainty equations:

U_{95} is identical to U_{ISO}.

$$U_{95} = U_{ISO} = \pm t_{95}[(b_R)^2 + (s_{\overline{X},R})^2]^{1/2} = \pm t_{95}[(U_A)^2 + (U_B)^2]^{1/2}$$

b_R = **systematic standard uncertainty**

$$b_R = \left[\sum\limits_{i=1}^{N} (b_i)^2\right]^{1/2}$$

$s_{\overline{X},R}$ = **random standard uncertainty**

$$s_{\overline{X},R} = \left[\sum\limits_{i=1}^{N} (s_{\overline{X},i})^2\right]^{1/2}$$

U_A = **ISO Type "A" standard uncertainty**

$$U_A = \pm\left[\sum\limits_{i=1}^{N} (U_{A_i})^2\right]^{1/2}$$

U_B = **ISO Type "B" standard uncertainty**

$$U_B = \pm\left[\sum\limits_{i=1}^{N} (U_{B_i})^2\right]^{1/2}$$

Alternate uncertainty models:

$$U_{ADD} = \pm[b_R + t_{95}s_{\overline{X},R}] \qquad U_{RSS} = \pm[(b_R)^2 + (t_{95}s_{\overline{X},R})^2]^{1/2}$$

U_{95} for the average of M results:

$$U_{95} = \pm t_{95}\left[\frac{\sum\limits_{i=1}^{N} (b_i)^2}{4} + \frac{\sum\limits_{i=1}^{N} (s_{\overline{X},i})^2}{M}\right]^{1/2}$$

Degrees of freedom for U_{95} equals Degrees of freedom for U_{ISO}:

$$d.f. = v = \frac{\left[\sum\limits_{i=1}^{N} (s_{\overline{X},i})^2 + \sum\limits_{j=1}^{M} (b_j)^2\right]^2}{\left[\sum\limits_{i=1}^{N} \frac{(s_{\overline{X},i})^4}{v_i} + \sum\limits_{j=1}^{M} \frac{(b_j)^4}{v_j}\right]} = \frac{\left[\sum\limits_{i=1}^{K} (U_{A_i})^2 + \sum\limits_{j=1}^{L} (U_{B_j})^2\right]^2}{\left[\sum\limits_{i=1}^{K} \frac{(U_{A_i})^4}{v_i} + \sum\limits_{j=1}^{L} \frac{(U_{B_j})^4}{v_j}\right]}$$

pooled standard deviation	$s_{X,\,pooled} = \left[\sum\limits_{i=1}^{N} v_i(s_{X,\,i})^2 / \sum\limits_{i=1}^{N} (v_i)\right]^{1/2}$

Calibration uncertainties are fossilized	$b_{CAL} = U_{cal} = \pm t_{95}[(b_{CAL})^2 + (s_{X,\,cal})^2 / N_{cal}]^{1/2}$

Equations for uncertainty (error) propagation

uncertainty (error) propagation for independent error sources:

$$s_{\overline{X},\,R} = \left[\sum_{i=1}^{N}\left(\frac{\partial F(X_{1,\,2,\,...i})}{\partial(X_i)}\right)^2 (s_{\overline{X},\,i})^2\right]^{1/2}$$

$$b_R = \left[\sum_{i=1}^{N}\left(\frac{\partial F(X_{1,\,2,\,...i})}{\partial(X_i)}\right)^2 (b_i)^2\right]^{1/2}$$

It should be noted that for the next three equations illustrating uncertainty propagation, the terms "U" with subscripts all represent uncertainties. That is, they are all either $s_{\overline{X}}$ values, s_X values, or b values. The extra terms needed for correlated errors or uncertainties are usually zero for random components and $s_{\overline{X}}$ and s_X. The value noted as U_{ij} remains the covariance of U_i on U_j.

uncertainty (error propagation) for non-independent error sources:

$$U = \pm\left[\sum_{i=1}^{N}\left(\frac{\partial F(X_{1,\,2,\,...i})}{\partial(X_i)}\right)^2 (U_i)^2 + \sum_{i=1}^{N-1}\left[\sum_{j=(i+1)}^{N}\left(\frac{\partial F}{\partial X_i}\right)\left(\frac{\partial F}{\partial X_j}\right)(r_{i,\,j})U_iU_j\right]\right]^{1/2}$$

$$(r_{i,j})U_iU_j = U_{ij} \text{ where } U_{ij} = \text{the covariance of } U_i \text{ on } U_j$$

Uncertainty propagation for two independent error sources, U_Y and U_Z, [the function is $X = f(Y,Z)$]:

$$U_X = \pm\left[\left(\frac{\partial f}{\partial y}\right)^2 (U_y)^2 + \left(\frac{\partial f}{\partial z}\right)^2 (U_z)^2\right]^{1/2}$$

Uncertainty propagation for two non-independent error sources, U_Y and U_Z, [the function is: $X = f(Y,Z)$]:

$$U_X = \pm\left[\left(\frac{\partial f}{\partial y}\right)^2 (U_y)^2 + \left(\frac{\partial f}{\partial z}\right)^2 (U_z)^2 + 2\left(\frac{\partial f}{\partial y}\right)\left(\frac{\partial f}{\partial z}\right)r_{y,\,z}U_yU_z\right]^{1/2}$$

$$r_{y,\,z} = \frac{N\sum y_iz_i - (\sum y_i)(\sum z_i)}{[(N\sum(y_i)^2 - (\sum y_i)^2)(N\sum(z_i)^2 - (\sum z_i)^2)]^{1/2}}$$

This reduces to the following for systematic uncertainties where the prime (') indicates the correlated elemental systematic uncertainties:

$$b_X = \left[\left(\frac{\partial f}{\partial y}\right)^2 (b_y)^2 + \left(\frac{\partial f}{\partial z}\right)^2 (b_z)^2 + 2\left(\frac{\partial f}{\partial y}\right)\left(\frac{\partial f}{\partial z}\right)b_y'b_z'\right]^{1/2}$$

General expression:

$$b_R = \left[\sum_{i=1}^{N}\left\{[(\theta_i)^2(b_i)^2] + \left[\sum_{j=1}^{N}\theta_i\theta_jb_{i,\,j}(1-\delta_{i,\,j})\right]\right\}\right]^{1/2}$$

$$b_{i,\,j} = \sum_{l=1}^{M} b_i(l)b_j(l)$$

To calculate the systematic uncertainty for an error source in both X and Y use:	$b_{YNET,i} = b_i(dY/di) - (b_i)(dX/di)(dY/dX)$

Equations to use for weighted uncertainty:

$$\bar{X} = \sum_{i=1}^{N} W_i \bar{X}_i$$

$$W_2 = (u_1)^2 / [(u_1)^2 + (u_2)^2] = 1 - W_1$$

$$W_i = \frac{\left(\dfrac{1}{u_i}\right)^2}{\displaystyle\sum_{i=1}^{N} \left(\dfrac{1}{u_i}\right)^2}$$

$$\bar{\bar{b}} = (W_1^2 \theta_1^2 b_1^2 + W_2^2 \theta_2^2 b_2^2)^{1/2}$$

$$\bar{\bar{s}}_{\bar{X}} = (W_1^2 \theta_1^2 s_{\bar{X},1}^2 + W_2^2 \theta_2^2 s_{\bar{X},2}^2)^{1/2}$$

$$d.f. = \bar{\bar{v}} = \frac{\left((W_1 \theta_1 s_{\bar{X},1})^2 + (W_2 \theta_2 s_{\bar{X},2})^2 + (W_1 \theta_1 b_1)^2 + (W_2 \theta_2 b_2)^2\right)^2}{\dfrac{(W_1 \theta_1 s_{\bar{X},1})^4}{v_{s\bar{X},1}} + \dfrac{(W_2 \theta_2 s_{\bar{X},2})^4}{v_{s\bar{X},2}} + \dfrac{(W_1 \theta_1 b_1)^4}{v_{b1}} + \dfrac{(W_1 \theta_1 b_2)^4}{v_{b2}}}$$

$$W_1 = \left(\frac{1}{u_1}\right)^2 / \left[\left(\frac{1}{u_1}\right)^2 + \left(\frac{1}{u_2}\right)^2\right]$$

Weighted uncertainty models:

$$\bar{\bar{U}}_{95} = \pm \bar{\bar{2}}[(\bar{\bar{b}})^2 + (\bar{\bar{s}}_{\bar{X}})^2]^{1/2}$$

$$\bar{\bar{U}}_{ADD} = \pm(\bar{\bar{b}} + t\, \bar{\bar{s}}_{\bar{X}})$$

$$\bar{\bar{U}}_{RSS} = \pm(\bar{\bar{b}}^2 + (t\bar{\bar{s}}_{\bar{X}})^2)^{1/2}$$

To determine the **significance of a correlation coefficient, r,** compare this t_{95} to Student's t_{95}:

$$t_{95} = \frac{|r|(N-2)^{1/2}}{(1-r^2)^{1/2}} \quad \text{for } N \text{ data pairs with } (N-2) \text{ degrees of freedom}$$

To calculate **median ranks** for probability plotting use **Benard's formula:**

$$P_{50\%} = \frac{i - 0.3}{N + 0.4} \times 100$$